THE CHARLES ASGILL AFFAIR

SETTING THE RECORD STRAIGHT

HERITAGE BOOKS
2024

HERITAGE BOOKS
AN IMPRINT OF HERITAGE BOOKS, INC.

Books, CDs, and more—Worldwide

For our listing of thousands of titles see our website
at
www.HeritageBooks.com

Published 2024 by
HERITAGE BOOKS, INC.
Publishing Division
5810 Ruatan Street
Berwyn Heights, MD 20740

Copyright © 2023 Anne Ammundsen

All rights reserved. No part of this book may be reproduced or transmitted in any form or by any means, electronic or mechanical, including photocopying, recording or by any information storage and retrieval system without written permission from the author, except for the inclusion of brief quotations in a review.

International Standard Book Number
Paperbound: 978-0-7884-2908-8

Graeme

In gratitude for all your help

CONTENTS

Illustrations		vii
Foreword		xi
Acknowledgments		xiii
Introduction		xvii

Part I

Chapter One	The Drawing of Lots	1
Chapter Two	Early Life and Education	6
Chapter Three	American War of Independence	18
Chapter Four	Capitulation at Yorktown	23
Chapter Five	Captain William McCurdy	27
Chapter Six	Condemned to be Hanged	32
Chapter Seven	Under Close Arrest	44
Chapter Eight	Intervention from France	57
Chapter Nine	The Journey Home	67
Chapter Ten	Paris; to thank Their Majesties	77
Chapter Eleven	James Gordon; 3rd Laird of Ellon	83
Chapter Twelve	The False Rumours	91
Chapter Thirteen	Troubles in America Now Behind Him	101
Chapter Fourteen	Irish Uprising	114
Chapter Fifteen	On the General Staff in Dublin	124
Chapter Sixteen	Jemima Sophia (Ogle) Asgill	133
Chapter Seventeen	Retirement from the Army	139

Chapter Eighteen	Thomas Graham; Lord Lynedoch	151
Chapter Nineteen	The Swindler Asgill	161

Part II

Chapter Twenty	William Charles Asgill	173
Chapter Twenty-One	Charles Childs	180
Chapter Twenty-Two	Mary Ann (Goodchild) Mansel	191
Chapter Twenty-Three	The Journey of Discovery	196
Appendix One	Letters Omitted from Washington's Published Correspondence on Asgill's Confinement	205
Appendix Two	The Conduct of General Washington, respecting the Confinement of Capt. Asgill, placed in its true Point of Light	208
Appendix Three	Captain Charles Asgill's letter dated 20 December 1786	220
Appendix Four	An eye-witness account of the drawing of lots on 27 May 1782	230
Appendix Five	Lady Asgill's letter to the comte de Vergennes	235
Appendix Six	Jean-Louis Le Barbier's Play	237
Appendix Seven	Asgill in Fiction	244
Appendix Eight	Sophia Asgill's letter to Thomas Graham	252
Appendix Nine	"The Swindler Asgill" Press Reports	255
Bibliography		260
Notes		269
Index		293

ILLUSTRATIONS

Cover: Sir Charles Asgill, 2nd Baronet. Mezzotint by Charles Turner, after an 1822 portrait by Thomas Phillips, RA. Coloured by Alan Birch	
The Reverend Kristin Miles with Anne Ammundsen at Trinity Church, New York City, on 29 May 2019	xvi
The Asgill & Ogle Family Trees showing only the main characters, by Alan Birch	xviii
The Drawing of Lots at the Black Bear Tavern on 27 May 1782, by Alan Birch	3
Coat of Arms of Sir Charles Asgill, 2nd Baronet	4
Sir Charles Asgill, 1st Baronet. Copy of the original portrait (painted by Thomas Hudson c.1760) by an unknown artist	7
Lord Mayor's Golden Coach at the Museum of London	8
Sir Charles Asgill's Bank at 70 Lombard Street	9
Westminster School, where the Rev. Samuel Smith L.L.D. was Asgill junior's Headmaster	11
Autograph book signed by Charles Asgill at the University of Göttingen in 1778	12
29 Old Burlington Street, Asgill's first home at the age of sixteen	13
A Lieutenant and Captain of the 1st Regiment of Foot Guards	16
Article XIV of the Articles of Capitulation of the British at Yorktown	35
Market Square, Lancaster, Pennsylvania, in 1782, showing the location of the Black Bear tavern.	39
Colonel Elias Dayton's House in Chatham New Jersey	45

Timothy Day's Tavern, Chatham, New Jersey	47
A modern map of Chatham, New Jersey, showing the 1782 locations of Timothy Day's Tavern and Colonel Elias Dayton's house	49
General Sir Guy Carleton, 1st Baron Dorchester, C-in-C British Forces in North America in 1782	53
Richmond Place, (now known as Asgill House) from front garden, 4 June 2002	57
Drawing Room at Asgill House	58
Amelia Angelina (Asgill) Colvile, Asgill's eldest sister	59
Charles Gravier, comte de Vergennes, French Foreign Minister to King Louis XVI	61
No.1 The Broadway, Old Kennedy House, where the British Commander-in-Chief's Headquarters was located in 1782	71
The route covered by Asgill during his time in America 1781-2, by Alan Birch	72
Their Majesties Queen Marie Antoinette and King Louis XVI	80
Mary Ellen Gordon, James Gordon's sister	85
The Morris-Jumel Mansion in Upper Manhattan, New York City	86
The room in which James Gordon is believed to have died	88
A memorial stanchion erected at Trinity Church, New York City, in honour of Lt. Col. James Gordon	89
First page of Asgill's Letter of 20 December 1786	94
Extract from Washington's letter to David Humphreys, 26 December 1786	96
HRH Prince Frederick, Duke of York and Albany, to whom Asgill was an Equerry	102

Georgiana Cavendish, Duchess of Devonshire, a Whig friend of the Asgills	106
Charles Asgill's wife, Sophia, Lady Asgill, née Ogle	108
Sophia's half-sister, Barbarina Brand, Lady Dacre, née Ogle	118
The hot water urn presented to Charles Asgill by the people of Clonmel	122
Lady Asgill's Setter Dog	138
Entry to the State Apartments, Dublin Castle	140
Richard Brinsley Sheridan. Playwright, poet and Whig politician	145
Admiral Sir Chaloner Ogle, 1st Baronet. Sophia's father	146
Admiral Sir Charles Ogle, 2nd Baronet. Sophia's brother	149
General Sir Thomas Graham, 1st Baron Lynedoch	153
HM King George III of Great Britain and Ireland	174
Letter from Lucy and Mary Ann Mansel to Herbert Mansel, 29 May 1833	181
Hill House, Loose, believed to be Mary Ann Mansel's house	186
Mary Ann and Charles Childs share the same grave at All Saints Church, Loose, near Maidstone in Kent	188
Mary Ann Mansel's death certificate	193
William George Maton M.D. English physician	194
Asgill depicted in *Two Girls of Old New Jersey: A School-Girl Story of '76*	242

FOREWORD

The story of Captain Asgill might have come from the pages of an adventure novel. And he really did figure in an 18th-century play, written in France to honour his mother's courage. I will certainly not spoil it for the reader by summarizing even the main points of his story. Suffice it to say that George Washington, Marie Antoinette and Louis XVI all played important roles. Though Asgill was a young man caught up in great events by chance, his experiences are able to tell us much about the ethics of 18th century warfare, ideas of gallantry and civilized conduct, the complex relationships between the Americans, the British and the French, and the conflict of human emotions with the demands of realpolitik. But his adventures did not end with this early episode.

That this story can at last be fully told reflects great credit on its author, Anne Ammundsen. She has battled against all the odds to find out the full facts of the adventures of the young officer who happens to be her ancestor, showing a persistence that few historians can match, and detective skills that Scotland Yard might envy in tracking down previously unknown or long forgotten evidence. She has succeeded at last in writing an accessible account that will both benefit scholars and entertain readers.

Robert P. Tombs

British historian of France, and Professor Emeritus
of French History at the University of Cambridge
and a fellow of St John's College

ACKNOWLEDGEMENTS

I've had so much help from kind and interested people, some of whom have now passed away. First amongst them must be the late Janet Reakes, Genealogist, of the Hervey Bay Family History Association, Queensland, Australia. She showed me where to search and more besides, during an online Ancestry.com course. Without Janet I'd never have found the late Kathleen Luckcuck, the distant cousin I had never known, who told me of my connection to Charles Asgill. She provided me with all the basic information, which resulted in me being hooked by Asgill's story ever since. She also accompanied me on my first visit to Asgill House, in 2002, where the owner, Fred Hauptfuhrer, was most welcoming (he is an American; the former Editor of *People*, who restored Asgill House, in 1969, to its former glory). He had some astonishing news; he handed us copies of letters he had received from Charles Asgill's descendants. We were flabbergasted, since Asgill died without issue. It started a long trail, the end of which may never be reached.

My former boss at Arbuthnot Latham, a private bank in the City of London, the late Peter Drummond-Murray, Slains Pursuivant of Arms to the Earl of Errol, helped me so much in the very early days of my research. It was a joy to be back in touch with him too.

Professor Robert Tombs, historian, specialising in French history at Cambridge University, helped to make the impossible happen and my gratitude to him is beyond words to convey. I was bowled over when he agreed to write the foreword to this book.

My gratitude also goes to LancasterHistory, Lancaster, PA, for being brave enough to publish Asgill's recently-discovered letter, in 2019. How fitting that this was published in America, the destination to which Asgill had written. The Editor of Family Tree, Helen Tovey, enabled me to take LancasterHistory's research to the next level, when she interviewed me on 7 March 2022. I am so grateful to Helen for that leg-up.

The man I had believed to be another distant cousin, Andrew Morrison, was the man I turned to for all matters relating to the Asgill family, both ancient and modern. We worked together for 20 years. The Colvile family (descendants of Asgill's eldest sister Amelia) provided information and allowed me to photograph her portrait. I would never have known about this image without their input. They also provided me with a photograph of their portrait of the Lord Mayor, which is a copy of the original. The late Lieutenant Commander Richard Colvile, RN, was kindness itself when I visited him about a year before he died, and his son, Charles Colvile, provided the approval needed to publish Asgill's 1786 letter. My cousin, William Newton, kindly provided IT assistance, which had been entirely beyond me. Pam Moore helped with information on Sophia (Ogle) Asgill's family, the Ogles of Martyr Worthy, Hants. So did the late Margot Strickland, who knew so much about the fascinating lady who was Sophia's half-sister, Barbarina Brand. The late Robert Maxtone-Graham (a kinsman of Lord Lynedoch) was very helpful with regard to Lord Lynedoch and his relationship with the Ogle sisters.

I eventually tracked more descendants of the wider story than I can even count. Some were very interested; some provided snippets of information; some really weren't interested at all! If only the Childs Family Bible had been found, who knows where that might have taken us? I was astonished to learn that, had it been found, it would have been in my own home town! It was an interesting exercise tracking about 35 of them down. Archivists and historians in libraries, museums and history centres often went above and beyond my original query; none more so than Meg Bower, and Janet Portman at the British Library. I was always touched by how interested they became in Asgill's story.

I am grateful to Paul Lay, the former Editor of *History Today,* who published my first article, "Saving Captain Asgill," in their December 2011 edition. Another publication (*Metropolitan,* the journal of The London Westminster & Middlesex Family History Society) kindly published three more articles. To everyone who has backed me and helped me, I owe a great debt of gratitude. My husband, Graeme, has been an outstanding supporter and help to me, and his checking of endless transcriptions with me must be seared on both our memories, so numerous were my requests for double-checking the final transcription. When the never-before-known portrait of Sophia Asgill materialised, within a John Hoppner portrait of the Duchess of York and Albany, Deborah Gage sealed the proof by kindly providing me with an image of Sophia as a child, in the possession of Nicolas Gage, 8th Viscount Gage. My thanks must also extend to a Saudi prince who allowed me to view the original Hoppner portrait in his possession, which enabled me to get a wonderful photograph of Sophia.

I must also acknowledge the assistance I have received from Professor Gregory Urwin, of Temple University, Doylestown, Pennsylvania, USA; and Don N. Hagist, the Managing Editor of the *Journal of the American Revolution,* both of whom never failed to answer my numerous questions. In the case of Don Hagist, over a nearly 20-year period he went way above and beyond, and I owe him an enormous debt of gratitude. Patrick Kerwin at the Library of Congress has been little short of amazing in the information he has so generously passed to me. Alison Thurman, Correspondence Unit Head at the North Carolina State Archives is another who falls into the "amazing" category when it came to my search for John Gregory; she took a personal interest in being the one to find him! Linnea Bass has been so kind too; Linnea, Wm. Thomas Sherman, and Chris Noel worked hard on the "Gregory" conundrum. As did the Research and Instructional Services Department at the Wilson Special Collections Library of The University of North Carolina at Chapel Hill, who went out of their way to help me. The historian James Taylor came to my aid when the "Gregory" conundrum became unsolvable, but at the time of writing, remains so. And far from least, my gratitude must go to Robin Carroll-Mann of Summit Free Public Library, New Jersey, USA, without whom I would never have discovered the location of Asgill's close confinement. She even sourced a drawing of the building, from the kind people at the Chatham Historical Society, the copyright owners. That was a very emotional moment in my own story of researching Asgill. And talking of images, where would I have been without my lovely artist, Alan Birch. Thank you, Alan.

As will be discovered in the pages of this book, Lieutenant Colonel James Gordon, 80th Regiment of Foot (Royal Edinburgh Volunteers), was the hero of the Asgill Affair. I came to love him as much as Asgill did himself. Gordon's sister's descendant, Robert Balfour, the Lord Lieutenant of Fife, has been supportive of my wish to have Gordon honoured by a stanchion in his name erected at Trinity Church, Wall Street, Manhattan, USA. This was erected on 8 March 2022. I am enormously indebted to the Rev. Kristin Miles for all her efforts in this regard. It was a joy to meet her when I was in New York in 2019, and she very quickly understood what a special man Gordon was.

My journey was made more wonderful by Katherine Mayo, the author of *General Washington's Dilemma*, which is about the Asgill Affair. Her meticulous research was, for me, awe inspiring, and I derived enormous pleasure from treading in footsteps trodden by her nearly 85 years before me. Were she to be alive, I hope she would like this sequel.

The Reverend Kristin Miles with Anne Ammundsen at Trinity Church, New York City, 29 May 2019. Author's collection.

INTRODUCTION

After studying Charles Asgill for 21 years I think I can safely say that I am an authority on this man, even if on nothing else. Many hundreds of people have helped me and I couldn't possibly have learnt so much without their help.

So many unaccounted-for coincidences have happened during my studies of him, that I do believe I have had his blessing, his cooperation and his help. He has been silenced (I can only conclude deliberately) for 233 years but I think, in truth, he has been my "ghost-writer."

I would like this book to be an enjoyable read for all, no matter the level of expertise each reader may be coming from. I have brought in characters who I am sure were involved, even if I have no proof. The Lord Mayor's funeral, and the Asgill family visit to Paris, being examples of this. In fact, I think I have been quite restrained.

In essence the pages of this book cover true events, with some conjecture on my part, to link the known facts into a coherent and smooth-running story. I had no option over this, since I could not speak to Asgill and ask him to fill in the gaps for me. So, please accept it and enjoy this book for what it is; not, what it isn't. It is the story of one man's life, not an academic tome.

I've had a few articles published on this subject, both in the UK and the US, but I cannot claim to be an author. My army upbringing was that of a nomad and it played havoc with my education, so I cannot claim to be an academic either. I simply want to share my findings; that is all. I do not deny that there is a sprinkling of my own supposition, based on my military background, my diplomatic life with my now retired ambassador husband, and my knowledge of Asgill himself. This is my claim, nothing more. I know very little about the history of the battles Asgill fought, so I have used the works of other authors liberally in order to bridge several gaps in my knowledge.

Anne Ammundsen

Kent, United Kingdom

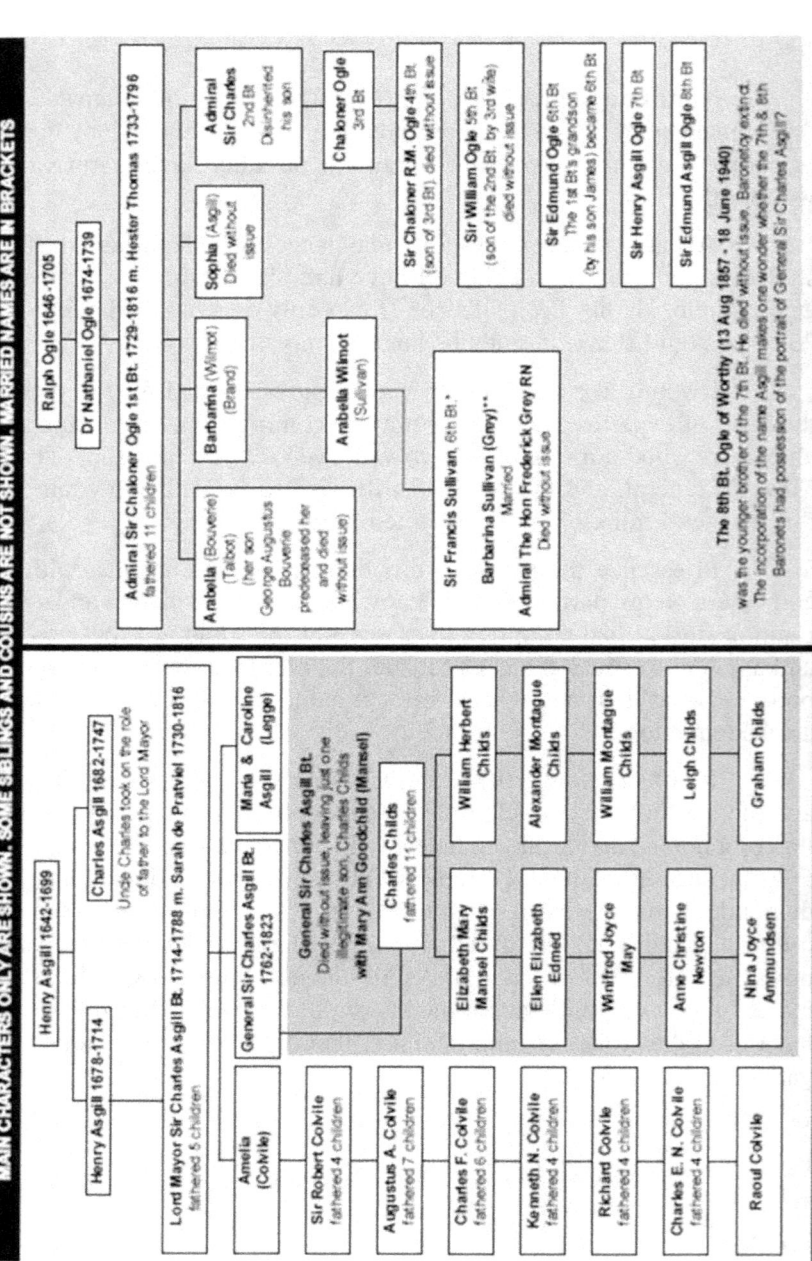

Illustration by Alan Birch

xviii

PART I

CHAPTER ONE
The drawing of Lots

The young drummer, beating his slow beat, didn't help matters at all. It might as well have been the knell of death. In fact, it was. With each beat, they all knew that the next piece of paper, drawn from the hat, might have their name on it; and they knew full well that the second hat, drawn from immediately afterwards, meant life or death if it brought forth the fatal "Unfortunate" lot. That man would go to the gallows. It was Monday, 27 May 1782 and thirteen British army officers, all known to one another, some of them dearest friends, and all holding the rank of captain, had been ushered into a small upstairs room at the Black Bear Tavern in Lancaster, Pennsylvania.

They sat in a circle waiting to know which of them would go to the gallows. Brigadier General Moses Hazen, the Continental Army officer in charge of the prisoner-of-war camp in Lancaster, to whom General George Washington had delegated the task of selecting which man was to die, performed his duty with the greatest delicacy, making it very clear that he abhorred his task, saying that he was but a servant and had to abide by his master's orders. Before proceedings began he had implored the British officers to pick the pieces of paper out of the hats themselves, but naturally they refused to take any part in this violation of a solemn treaty which George Washington himself had signed seven months before.[1] When these officers had surrendered themselves as prisoners-of-war at Yorktown, Virginia, in October 1781, the Articles of Capitulation included specific wording to prevent the exact scenario that they were now in: "No article of capitulation to be infringed on pretence of reprisals."[2]

And yet here they were. This drawing of lots was a violation of a clause in a solemn treaty, a clause which Washington had conveniently disregarded in his decision to take revenge for the murder of an American militia captain – a murder committed by irregular forces which were not part of the British regular army nor in any way associated with the Yorktown prisoners.

Tit-for-tat murders between factions for and against American independence had been going on for a long time prior to the event in which these officers now found themselves entwined. The American Revolution is known as The American War of Independence on the British side of the Atlantic, but this war had been a Civil War too. Not all Americans had wished to sever their ties with king and country. Many thousands of American colonists, known as Tories, Loyalists or Refugees, had remained loyal to King George III. Even though there had been countless savage acts of bloodshed and retribution during seven years of war, the recent death of an American militia captain at the hands of Loyalist militia regulars had sparked demands for revenge, demands to which Washington acquiesced by allowing one captive officer to be selected for execution. This meant thirteen young British army officers now sat waiting to know which of them would die. As each draw took place, the tension grew with each drum beat. Ten of the friends now knew they would live and their main focus was dread for their remaining three friends, whose names still had to be drawn: Charles Asgill, George Ludlow, and one other remained uncalled.

The drum beat its knell. Name number eleven was drawn from the hat. A pause, before another drummer picked the accompanying piece of paper, which would either be another blank, or it would be the "Unfortunate" lot. The American Commissary of Prisoners read the name.[3]

Lieutenant and Captain Charles Asgill of the First Regiment of Foot Guards was called.

Another pause; another drum beat, this time making his heart beat much faster than the drum.

The Commissary of Prisoners read the piece of paper drawn after the name: "Unfortunate."

Captain Asgill had prepared himself mentally for this. They all had. He knew he had the resolve to face death, however unjustified that death would be. But the human spirit doesn't always give one the strength needed, not in such an outrageously bizarre situation. He must have been utterly ashamed that his reaction was to burst into tears. In unison, they all did. One of Asgill's dearest friends, Henry Greville, wrote to his mother a couple of days later: "I cannot think the most bloody Field of Battle could wear half the terrors this hour did … There were more tears shed here the 27th May than ever fell on any occasion."[4]

When Asgill had recovered his senses, he exclaimed "I knew how it would be; I never won so much as a hit at backgammon in my life!" He added, "the lot fell to me and I will abide by it."[5]

The Drawing of Lots

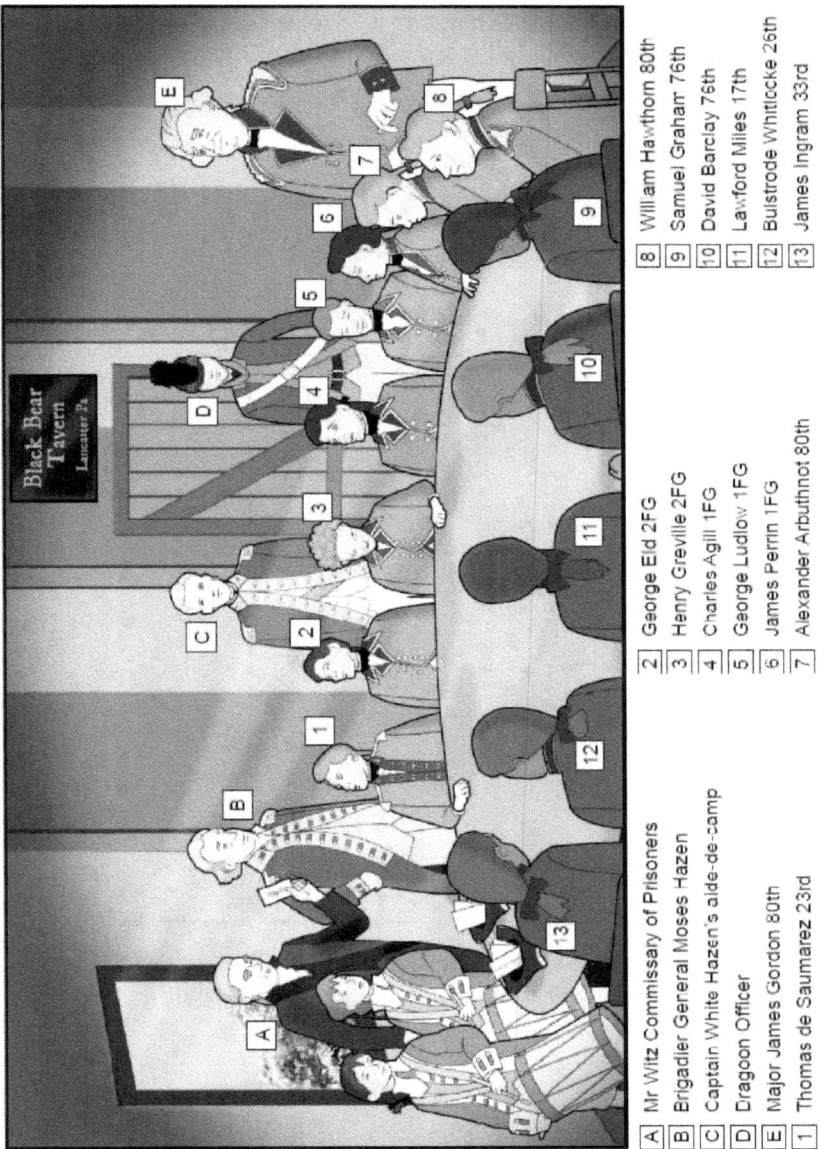

The Drawing of Lots at the Black Bear Tavern, 27 May 1782. An artist's impression drawn from archive records. By Alan Birch 2022. Copyright held by the author.

The Charles Asgill Affair

Asgill's family motto must have come to his mind: *Sui oblitus commodi* – "regardless of his own interest." He needed to live up to it now, more than at any other time in his short 20 years of life.

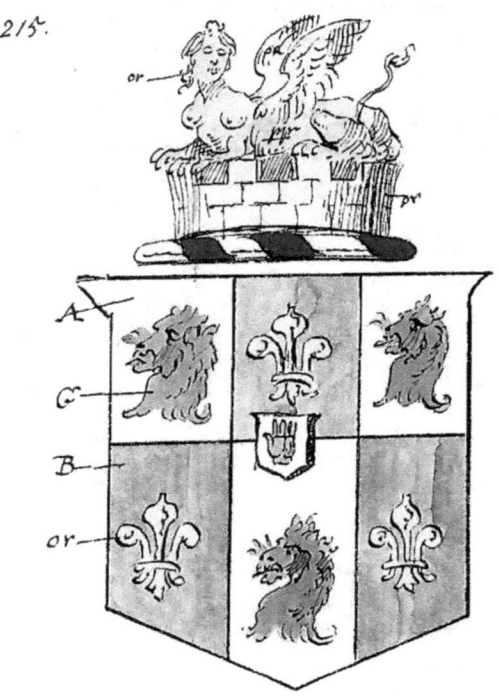

Coat of Arms of Sir Charles Asgill, 2nd Baronet. Produced and supplied on demand by the College of Arms, on request. Copyright passed to the author on purchase.

The townsfolk of Lancaster had watched in amazement as the captive British officers filed into the Black Bear Tavern, laughing and talking animatedly with one another. When they came out, everyone was in tears, except, by then, Asgill. Their commanding officer, Major James Gordon of the 80th Regiment of Foot (Royal Edinburgh Volunteers) had jumped to Asgill's side when his name was called and had taken hold of

The Drawing of Lots

him, snapping in his ears, "For God's sake, don't disgrace your colours!" It was just the tough love he had needed and brought him to his senses.[6]

Henry Greville's letter to his mother, continued: [7]

> he is an amiable young man perfectly adored by his Parents, esteemed and respected by his friends. When the Commissary of prisoners mentioned the name of the Sufferer he burst into tears, and for some time could not articulate, we were all indeed very much agitated, Major Gordon in particular, never did a man possess more exquisite sensibility than he does; he is an honour to Society, and his attention and care respecting everything that could alleviate my poor Friend's misfortune will ever be remembered with Gratitude and esteem by his Brother Officers. Asgylle tho' evidently affected with his Fate, yet bore it with uncommon resolution, he thanked Genl Hazen in the prettyest style for the politeness he had shown him, and took his leave. I walked home with him and endeavoured to keep up his spirits, at first he was melancholy and thoughtfull, seemingly very unhappy, but it gradually wore off, and no person to have been in his company could have supposed he was doomed to die.

CHAPTER TWO
Early life and Education

Charles Asgill was born in London on 6 April 1762 to a wealthy banker father, Sir Charles Asgill 1st Baronet, and his wife Sarah.[8] His first wife, Elizabeth Vanderstegen, had died after giving birth to their only child, a daughter, Annabella, on 11 January 1754. The 1st Baronet married Sarah Theresa Pratviel (12 February 1730 - 6 June 1816) on 12 December 1755 and by October of the following year a daughter, Lydia, had been born. Another daughter, Amelia, followed in 1757, but both lost their half-sister when Annabella died on 30 August 1761. The family continued to grow when Charles Jnr was born in 1762, then Maria in 1767, and Caroline in 1774. By this time Lady Sarah was 44 and Sir Charles was 60 years of age, but young Charles was their only son. This was a close-knit family and they enjoyed close bonds with one another. They never wanted for anything, since they were well provided for by a wealthy father. They all grew up with Whig Protestant values drummed into them; they were taught to always look after those less well off than themselves, and to seek the best ways of achieving social progress. They could afford to be charitable, in mind, spirit and good deeds, and it became part of the children's way of life through to adulthood.

The elder Charles, born on 17 March 1714, was a stern disciplinarian with a heart of gold. He was not easily manipulated, but was known to give in to his daughters' wiles. A man with a social conscience; traits he may have learned from his uncle Charles Asgill (1682-1747), who acted *in loco parentis* from his birth. Uncle Charles (who died childless) was the youngest brother of his father, Henry Asgill (1678-1714), who had died eleven days after his son's birth.

He started his career "at a banking house in Lombard street, as out-door collecting clerk." At this firm, William Pepys & Company, "he progressively rose by his merit to the first department in the house."[9] In 1740 he partnered with an established goldsmith, Joseph Vere, to form the banking and goldsmithing company Vere & Asgill. The Lombard Street firm thrived over the next decade; by 1750 it was styled Vere, Asgill and Co. The bank operated, as all banks did during that era, from private residences. That changed in 1757 when Asgill commissioned the architect Sir Robert Taylor to design the first purpose-built bank in the city, built at 70 Lombard Street, which survived until its demolition in 1915.[10] The

Early Life and Education

partnership continued to evolve, as Sir Charles Asgill, Nightingale & Wickenden in 1765 and Asgill, Nightingale & Nightingale in 1775.

Sir Charles Asgill, 1st Baronet (17 March 1714 – 15 September 1788). Copy of the original portrait (painted by Thomas Hudson c.1760) artist unknown. By kind permission of the Colvile family.

The Charles Asgill Affair

Success in business brought with it a knighthood in 1756, and a host of civic activities. Charles Asgill was known for his charitable work, such as service on the board of governors for Bridewell Royal Hospital and for St. Thomas' Hospital in Southwark.[11] Politically, he served as Alderman of Candlewick Ward for 22 years beginning in 1749, as sheriff of the City of London in 1753, and, most prominently, as Lord Mayor of London in 1757 and 1758.[12] The holder of the office of Lord Mayor is accorded precedence over all personages with the exception of the monarch (but only within the City of London). Asgill commissioned the Golden Coach, designed by the same Sir Robert Taylor that designed the bank, which is still in use today by the Lord Mayor during his annual November Lord Mayor's Show, a pageant that has survived over eight hundred years of London history including plagues, fires, floods and bombs. This is the second oldest working golden carriage in the country; The Speaker's State Coach is older, but the monarch's was built after both of these.

Lord Mayor's Golden Coach at the Museum of London, 11 January 2012. Author's collection.

Early Life and Education

Charles Senior owned several properties in London and farms in Essex, but the family lived in London, mainly at Portman Square, a home also designed by Taylor and described in 1812 as "one of the prettiest he ever built."[13] Taylor was not only a brilliant designer and architect, he was a close friend of Sir Charles, and was an integral part of the family as the children grew up. Were it possible to find out, it would not surprise me one bit to discover that Taylor was a godfather to one of the Lord Mayor's children. London was central to their lives, in order that their father could pursue his banking business.

Sir Charles Asgill's Bank at 70 Lombard Street, designed by Sir Robert Taylor in 1757 and demolished in 1915. Artist unknown. Wikimedia Commons

The Charles Asgill Affair

During 1772 and 1773, 30 banks across Europe failed in what came to be known as the British Credit Crisis of 1772-1773.[14] The *Gentleman's Magazine* stated that "No event for 50 years past has been remembered to have given so fatal a blow both to trade and public credit."[15] It was only through good stewardship that Sir Charles' bank survived, since banks that had never engaged in speculation did not bear any losses and gained prestige for their outstanding performance despite the turbulence.[16] But the banker had not been happy with the manner in which some of his colleagues in the banking world exacerbated the crisis. The British government was forced to introduce controversial taxation legislation for the American colonies in an attempt to remedy the crisis, which led to the Boston Tea Party in December 1773. That was a major catalyst leading to the American War of Independence. The 59-year-old banker Charles Asgill could not know that this sequence of economic events would eventually lead to his son's life hanging in the balance a decade later; he was, however, surely aware of a dance tune written to acknowledge his displeasure with his fellow bankers – it was aptly titled *Asgill's Rant*.[17]

The elder Sir Charles died in 1788. His obituary in the *Gentleman's Magazine* stated that "he was a strong instance of what may be effected even by moderate abilities, when united with strict integrity, industry and irreproachable character."[18] A hard-working and principled banker, he was often absent from home. So it was inevitably his wife who cared for the children, making sure they worked, played and worshiped. She was from a family of French Huguenots who had fled religious persecution in France and her faith was very important to her. Monsieur Elie Brilly was the French Huguenot Pastor of St. Jean (French Church), John Street, Spitalfields, and he was a close confidante of Lady Sarah, who was a regular attendee at his church. She had many friends who shared her beliefs, but she was also interested in politics, the arts, music and theatre. She was known as always kind and always fair in her dealings with people, and she taught her family well.

Of course her only son's education was particularly important to both her and her husband, so just before Charles Jnr's seventh birthday, on 29 March 1769, he was sent off to Westminster School as a boarder.[19] Charles Snr had been a student at Westminster himself, having enrolled when he was twelve years of age, in June 1726, where his Whig values were instilled in him.[20]

Early Life and Education

Westminster School, where Samuel Smith was Asgill junior's Headmaster. Wikimedia Commons.

The still six year old Charles must have missed his home, family, and the life they had enjoyed together, but soon made good friends at school. Some of them were American and good bonds were made there. Two would come to his aid later in life, Alexander Garden and Henry Middleton, and they were at school together from 1771 to 1775. Together, Garden and Middleton later jointly composed the following letter (written by Garden) about their school-friend, saying:

The Charles Asgill Affair

I had been a school-fellow of Sir Charles Asgill, an inmate of the same boarding-house for several years, and a disposition more mild, gentle, and affectionate I never met with. I considered him as possessed of that high sense of honour which characterizes the youth of Westminster in a pre-eminent degree.[21]

Young Asgill's parents would have ensured that he worked hard to gain a good education at Westminster School, but the days were long and the learning relentless. He might have hated it all at times, but would have known that this education would stand him in good stead for the rest of his life. Discipline was strict, but that too was to benefit him later. When school was over he went to Germany to attend the University of Göttingen, where he flourished.[22] On leaving three years later he signed a friend's autograph book with a quotation from Alexander Pope, writing, "An Honest Man is the noblest work of God" – trite perhaps, but he believed the sentiments conveyed.[23]

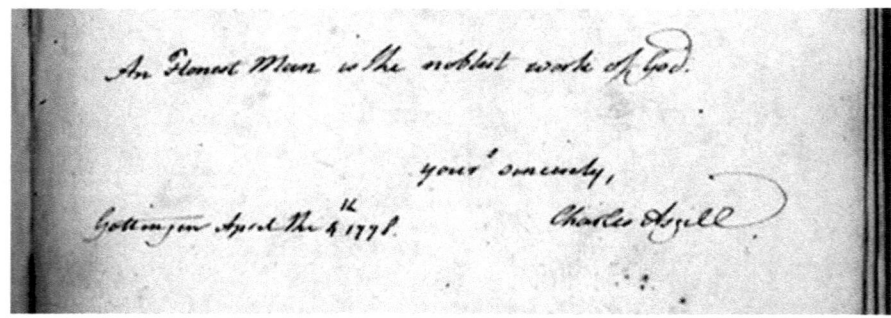

"An Honest Man is the noblest work of God." Autograph book signed by Charles Asgill at the University of Göttingen, 4 April 1778. Wikimedia Commons.

To his father's great disappointment, he became increasingly aware that his son would not be following him into the banking business, so he would be unable to keep the name of Asgill as a well-respected name in the City of London. On leaving Germany the young man was able to speak both French and German (the latter he eventually forgot, through lack of use).[24] On returning from Germany, at the age of 16, he moved into a house owned by his father at No. 29 Old Burlington Street. From 1778 until 1785 this was his home; although he was away for some of that time. It was all part of his father's plan to persuade him to become a resident of

Early Life and Education

London, since he had so cherished the hope that he would be following the banker's path into a career in business, but the son had other aspirations.

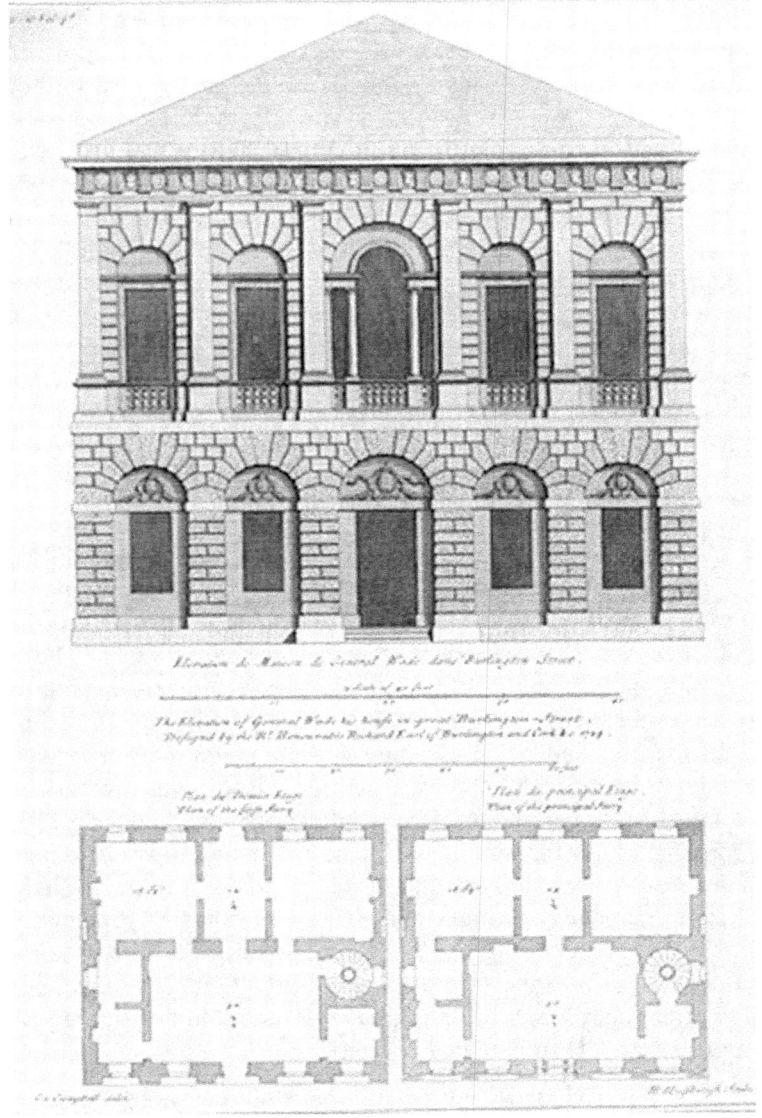

29 Old Burlington Street, (1778-1785), Asgill's first home. Etching and engraving by the Dutch-British printmaker Henry Hulsberg. Public domain.

The Charles Asgill Affair

The house was built in 1723 for General (later Field-Marshall) George Wade, and featured a highly-decorated, Italian-inspired exterior described by an observer in 1734 as "one continued cluster of ornament." The interior, on the other hand, was described by Horace Walpole in 1748 as "worse contrived on the inside than is conceivable, all to humour the beauty of the front." With an east-facing street front thirteen feet back from the road, it had on the ground floor a central hall twelve feet wide and eighteen feet deep, flanked by two eighteen-foot-square rooms; behind these was a large hall 30 feet wide and 20 feet deep with small, closet-like rooms on each side, all facing the garden. The upper floor followed the same plan. By the standards of a wealthy London family the house was small, but interesting, and today would be described as the young man's first foot on the property ladder, provided by the bank of Mum and Dad. Its Doric features may have suited young Charles in a way unintended by his father, for, in the words of a 1727 visitor, the former field-marshall's residence was well-suited for a man with military aspirations "because antiently Temples and monuments consecrated to Heroes were of this order."[25]

What young Asgill really wanted was to join the army. This was always going to be a problem for his parents, his father in particular. When Asgill moved into the Field-Marshall Wade's house, Britain had been at war with her rebelling American colonies for three years. The Asgill parents both had staunch Whig beliefs, and to fight the colonists in America was anathema to them; moreover, loving him as they did, they feared for his life. They tried pleading with their son, but to no avail. His father thought his bribe of a house would be failsafe, but it wasn't. So, he offered him £3,000 per annum if he would adopt some other profession.[26] The young man knew he was in trouble now, because how could he refuse such a generous offer? He turned to his eldest sister, Amelia, for help. She knew how to manage their Papa, and she did. Eventually, reluctantly, and with a feeling of foreboding, the father finally allowed his son to join the army as an ensign with the 1st Regiment of Foot Guards,[27] now known as the Grenadier Guards. His life's journey was about to begin; the year was 1778 and he was still fifteen years of age.

The Foot Guards were the monarch's household troops. Responsible for protecting royal properties, they were based in London and normally did all of their duties there, making them a highly desirable organisation for men with civic and political, as well as military, aspirations. But during times of war, detachments were often sent to the

Early Life and Education

theatre of conflict where, as well-trained and highly disciplined soldiers, they fought in the forefront of major campaigns. In 1776 a detachment of 35 officers and 1,100 soldiers from the three regiments of Foot Guards sailed to fight in the American War. Operating in America as the Brigade of Guards, they were in the thick of several major campaigns and suffered many casualties; replacements were sent on a regular basis. Dangerous though this service was, it was attractive to aspiring young officers anxious to prove their ability and bravery.

There was no formalised training for young army officers during this era. A newly commissioned ensign learned from his superiors. In the words of a textbook for young officers, "you should certainly acquit yourself with a scrupulous exactness of your several duties; and by your acquiring a habit of them, as you rise in rank, every duty will become easy to you."[28] "The first thing necessary for you to learn," the book continued, "in order to move with grace and exactness, is the military step (which will set you more upright than a dancing-master)." Referring to the flintlock muskets and other weapons used by the army, the book directed, "the use of the firelock, in every position, manual exercise, &c. should be your next consideration; then the salutes of the fuzee and espontoon, with the different facings practised by those weapons." But "As to the other parts of necessary duties, the adjutant will instruct you to them, and present you with a company's book of standing orders, forms of returns, &c."[29]

The initial expenses for a newly-commissioned officer were considerable. Asgill must have visited a tailor (almost certainly Gieves at 1 Savile Row) to bespeak uniforms, including a scarlet dress coat with royal blue lapels, cuffs and collar, all trimmed with gold lace and gold buttons. At least two more suits of clothing, two hats with cockades, a pair of leather gloves, a regimental sash and gorget, a sword, sword belt and sword knot, a pair of boots, a cloak, an ample supply of shirts, stockings, handkerchiefs, shoes and an assortment of towels, sheets, pillow cases and other necessities were among the goods required for a comfortable garrison life. Also required were administrative essentials like a writing case stocked with pens, ink, paper and sealing wax; drawing instruments; military textbooks; and "a watch, that he may mark the hour exactly when he sends any report."[30] Even though a gentleman like Asgill may have already owned some of these things, the bill for outfitting him in his new career was surely considerable.

A Lieutenant and Captain of the 1st Foot Guards, painted by Sir Joshua Reynolds in 1782. The Metropolitan Museum of Art, New York.

An ensign in the Foot Guards was paid at a rate of 5 Shillings 10 pence per day;[31] from this he paid for his own food and drink, washing of

Early Life and Education

his clothing and bedding, and consumables including "hair powder, pomatum, soap, black-ball, pens, ink, wax and wafers." He was allowed to engage a private soldier as a servant, paying him a shilling a week for services that included keeping his clothing and shoes in good order, helping him dress his hair and shave, running errands, and whatever else the officer might require. By one writer's estimate these expenses could amount to well over 3 Shillings per day, much more in metropolitan London.[32]

Two years after joining the army, on 10 December 1780 Charles Asgill was ordered to join the Brigade of Guards in America. While he was preparing for the journey, in February 1781 he became a lieutenant in the 1st Foot Guards; because of their standing as household troops of the monarch, officers in the Guards held higher ranks in the regular army, so Asgill's lieutenancy in the Guards made him also a captain in the army as a whole. He was still only 18 years of age. His father's fears about his son going to war became a reality. In the spring of 1781 young Asgill was sailing to America to join an army commanded by General Sir Charles, Earl Cornwallis, operating in the southern American colonies. The family parting must have been heartbreaking for them all. But young Asgill was overjoyed. He was doing what he had wanted to do for so long.

CHAPTER THREE

At War in America

Having received his marching orders in December 1780, Asgill went out with replacements for the Brigade of Guards in the spring of 1781. With him were Lieutenant Colonel Gerard Lake, and his friends Lieutenant and Captain Hon. George Ludlow and Lieutenant and Captain James Perrin, all of them in the 1st Foot Guards.[33] From London they went to Cork, Ireland, where they embarked on ships that would cross the Atlantic as a convoy. On 20 February Lake and Ludlow boarded the Royal Navy 50-gun warship HMS *Warwick*, which was under the command of the 35-year-old Captain George Elphinstone, 1st Viscount Keith.[34] At around the same time Asgill and Perrin apparently boarded transports, privately-owned vessels operating under contract to the royal navy, each carrying anywhere from a few dozen to a few hundred troops depending on the ship's size.

Young officers like Asgill supervised the soldiers carried on their individual transports, maintaining discipline and attending to nutrition and hygiene during the voyage. "Take Care that every Morning all the Men be brought upon Deck, the Births cleaned, and the Bedding brought up to air, if the Weather will permit," read typical orders to officers supervising soldiers on a voyage. "Allow no smoking below decks, and forbid gambling games and the selling of liquor. Bring the men on deck whenever possible for air and exercise. Sweep the berthing areas below decks regularly, and sprinkle them with vinegar for sanitation". Discipline was as important at sea as it was on land.[35]

After waiting for favourable winds the convoy sailed in late March, and arrived in Charlestown (today Charleston), South Carolina, on 3 June.[36] There some troops disembarked as reinforcements for that colony.[37] Then the *Warwick*, plus 15 victuallers and others, went on to Hampton Roads, Virginia at the mouth of Chesapeake Bay. Asgill and his comrades disembarked on 19 June, probably at Norfolk or Portsmouth.[38] The ship then continued its journey to New York.[39]

The army that Asgill joined, commanded by General Charles, Earl Cornwallis, was operating in Virginia at the mouth of Chesapeake Bay, where the bay, the James River and the York River all meet and empty into the Atlantic Ocean. The towns of Portsmouth and Norfolk were on the south shore; across the water to the north, Yorktown and Williamsburg were on a peninsula formed by the James and York rivers. The British forces there consisted of around 10,000 men including British regulars, German ("Hessian") auxiliaries and Loyalist regiments (troops raised in America for war); accompanying it were a few thousand civilians including artificers, soldiers' wives and children, refugees, escaped slaves and other followers.

About half of the army had come to Virginia from New York just a few months earlier, attempting to secure the region as part of an evolving – and failing – British strategy to pacify the American southern colonies. The springtime operations in Virginia were typical of the British army's experience in the war so far: local successes on the battlefield that did not produce any long-term strategic benefit.

The other portion of the army, including the Brigade of Guards, had been in the southern colonies for over a year, but had arrived in Virginia in late May. Led in person by General Cornwallis, they had fought a campaign in South Carolina in 1780, followed in 1781 by an arduous, circuitous march of over six hundred miles through the interior of North Carolina, attempting to catch and eradicate an elusive opponent. The campaign culminated in the Battle of Guilford Courthouse on 15 March, a pyrrhic victory in which Cornwallis's army, already depleted by illness and the rigors of spartan campaigning, suffered casualties that left it operationally ineffective. Cornwallis retreated to Wilmington, North Carolina, on the coast before taking his battered army north to join the British force in Virginia. He wrote,

> The conduct and actions of the officers and soldiers that composed this little army will do more justice to their merit than I can by words. Their persevering intrepidity in action, their invincible patience in the hardships and fatigues of a march of above 600 miles, in which they have forded several large rivers and numberless creeks, many of which would be reckoned large rivers in any other country in the world, without tents or covering against the climate, and often without provisions, will sufficiently manifest their ardent zeal for the honour and interests of their Sovereign and their country.[40]

The Brigade of Guards had suffered terribly in the 15 March battle. Their original 1776 strength of 35 officers and over 1,000 men had been reduced during five years of war, in spite of regular replenishment, to only about 19 officers and 450 men available to go into the battle; others were ill, infirm or recuperating in various garrisons and posts. The depleted brigade lost half its remaining strength in the battle, losing 11 officers and 200 soldiers killed and wounded. Those too severely wounded to move were left behind to become prisoners of war when the army withdrew to Wilmington. A month after the battle the brigade's commander, Brigadier General Charles O'Hara, described the men as "very Shatter'd, exhausted, ragged Troops."[41] Some had been fighting in America since 1776, and their war was not yet over.

As if warfare was not enough, the region's summer heat brought its own suffering. A German officer wrote, "for six weeks the heat has been so unbearable that many men have been lost by sunstroke or their reason has been impaired. Everything that one has on his body is soaked as with water from the constant perspiration. The nights are especially terrible, when there is so little air that one can scarcely breathe."[42] There was also "the torment of several billions of insects, which plagued us day and night."[43] These billions of insects brought malaria that incapacitated large numbers of troops throughout the summer.

Sparse records make it impossible to know exactly where Charles Asgill was during the summer months of 1781. He probably disembarked at Portsmouth, Virginia, on 17 June. On 24 June the officer commanding that garrison wrote to General Cornwallis in Williamsburg, informing him that he had sent personnel of the Guards from Portsmouth to Williamsburg;[44] by 26 June those guardsmen had arrived in Williamsburg.[45]

Detachments of the Brigade of Guards and a few other British and Loyalist regiments totalling some 300 men set out from Kemp's Landing, a river head near Norfolk, on the evening of 24 June. They arrived at Great Bridge, a few miles southwest, the following morning and prepared for an arduous march that would allow them to surprise a rebel post at the North Carolina border. Getting there meant crossing the Great Dismal Swamp – also called the Black Swamp. They set out in the evening; by two in the morning they were deep in the swamp. "For a mile every step we were two feet depth & some times three," wrote an officer of the slog. Every 30 men carried one light to guide them through the black waters. After three hours of trudging in the darkness they reached a river. A crude raft allowed six men at a time to cross, a tedious process that was nonetheless completed before 11 in the morning of 26 June. This put them close to their quarry, and so far they had remained undiscovered.[46]

At about 1 o'clock in the afternoon the advance guard, consisting of 20 men of the Brigade of Guards, a similar number of light infantry and about 50 Loyalist militiamen, encountered and pursued a party of the enemy, capturing a colonel, a captain and about 30 soldiers. They then found and burned a fort "at great swamp in north Carolina," and the next day destroyed works at "nor west landing," displacing rebel forces commanded by a North Carolina militia general named Isaac Gregory. The British force returned to Kemp's Landing by way of Great Bridge, completing the expedition by 28 June with the loss of only one man. The 26 June action is relatively obscure but has been referred to by historians as the Battle of Black Swamp.

Over a year later a Scottish publication, *The Hibernian Magazine*, published a story about Captain Asgill that included the enigmatic passage that he was "so well known to him by his bravery and humanity in different instances, particularly when the command devolving on him by the illness of his colonel, he took a post from the Americans, commanded by colonel Gregory, who being old and wounded, he supported him himself, with an awful and tender respect most filial, evincing the true greatness of his amiable mind."

This passage has been repeated by subsequent writers with no attempt to corroborate it.[47] Newspapers and news magazines of the era frequently published stories with inaccuracies, or sometimes that were entirely untrue. The only post taken from the Americans in the area where Asgill served was the fort and associated works at Black Swamp. General Gregory was not wounded in that battle; however, there was a *Colonel John Gregory* in the Camden County Regiment of the North Carolina Militia, but too little is known of him to determine whether he could have been the man indicated in the magazine account.[48] It is not clear whether the Guards detachment on the expedition was composed of the replacements that landed on 17 June, or soldiers already in the area. And no indication of Asgill's involvement, besides the magazine and other newspaper accounts, has come to light. If true, it reflects a compassionate character on Asgill's part, and also reflects the upbringing the Lord Mayor instilled in his children.

It is also possible that Asgill was not involved in the Black Swamp expedition. He may have been among the "recruits for the guards" sent from Portsmouth to Williamsburg between 24 and 26 June. It remains unknown when young Captain Asgill first saw action in America, but night time marches in the forests and swamps of Virginia were a far cry from royal parades in London and must have challenged the young officer's ability to adapt and endure.

The Charles Asgill Affair

Nearly 30 years later, in 1810, Asgill met up with another veteran of the Virginia campaign named Roger Lamb. Lamb enlisted in the British army in 1773, sailed to America with the 9th Regiment of Foot in 1776, became a prisoner of war in 1777, and escaped into British lines in New York in 1778. He joined the 23rd Regiment of Foot, attained the rank of sergeant, and served under General Cornwallis throughout the entire 1780 and 1781 campaigns. After the war he became a schoolmaster in Dublin, and in 1809 published a history of the American War that included many of his own experiences. He also repeats *The Hibernian Magazine* story and this probably brought him to the attention of Asgill, who met with him in Dublin.

They reminisced about their time in America. Asgill facetiously commented, "I will go to heaven for all the good actions I have done."[49] Lamb, by this time a very religious man who believed that the path to Heaven was through devotion to God rather than earthly deeds, was not impressed and perhaps with good reason thought Asgill to be conceited. But if *The Hibernian Magazine* and other press accounts are true, perhaps Asgill was justifiably proud of his noble action towards a wounded enemy officer.

CHAPTER FOUR
Capitulation at Yorktown

As 1781 progressed, strategic considerations overshadowed Cornwallis's tactical efforts to expand British holdings in Virginia. Combined American and French land forces put the vital British-held city of New York in danger. A large French naval force threatened the American coast. The sea was Cornwallis's lifeline for resupply, reinforcement or, if necessary, evacuation. In August he consolidated his army at Yorktown (often called simply "York" at the time) on the York River, a post that offered some security in that it could be fortified on the land side and had a sufficient anchorage on the river side. Across the river to the north, Gloucester Point was also secured. For a short while things appeared safe enough to allow the next strategic move to be determined.

Whatever Asgill's experiences had been in late June and July, August brought the tedious task of constructing defensive works around Yorktown and Gloucester. The days of officers like him were spent supervising teams of soldiers and labourers as they toiled to dig trenches and raise earthen walls, posting and monitoring guards and sentries, making sure soldiers received and prepared provisions and tended to hygiene and health, managing military paperwork, and countless other tasks. The attrition the army had suffered meant that every officer and soldier was pushed to the limits of their capacity in the hot summer sun punctuated by thunderstorms and lightning that was sometimes lethal.

The army's situation took a turn for the worse in early September. A rare French naval victory at the Battle of the Virginia Capes removed any prospect of the British army at Yorktown being relieved or reinforced from the sea. The American and French armies threatening New York made a masterfully quick move south, joining forces with American troops already in Virginia. As September drew to a close, Cornwallis's army was hemmed in, surrounded by a much larger army and blocked from escape by a powerful navy. The Americans and French closed in, building fortifications of their own, digging trenches ever nearer to the British lines, constructing batteries for cannons and mortars to bombard their opponents. The British returned fire with their own artillery, but it was not enough to turn the tide.

By the second week of October the situation in Yorktown was bleak. Supplies of every sort were running low. In the crowded confines of the British defences, smallpox broke out among the civilian labourers and followers of the army, many of them Blacks who had fled slavery for the freedom afforded by British service. Unable to endure much longer, there remained one option for the beleaguered army. On 16 October Cornwallis made a desperate attempt to save his army by escaping across the York River to Gloucester during the night. From there, an overland trek north offered some hope of eventually reaching New York, or some coastal place where the Royal Navy could meet them.

The desperate gamble did not go as planned. As Cornwallis described it:

> After making my arrangements with the utmost secrecy, the light infantry, greatest part of the guards, and part of the 23d regiment, landed at Gloucester; but at this critical moment, the weather, from being moderate and calm, changed to a violent storm of wind and rain, and drove all the boats, some of which had troops on board, down the river. It was soon evident, that the intended passage was impracticable; and the absence of the boats rendered it equally impossible to bring back the troops that had passed, which I had ordered about two in the morning.[50]

Almost a year later a British newspaper presented a dramatic account of this abortive escape attempt, praising the heroics of Charles Asgill. Like many newspaper accounts of the era it is not corroborated by other sources, leaving some doubt as to its accuracy:

> Captain Asgill ... had the command of the troops when Lord Cornwallis endeavoured to save part of his army by forcing them away from Gloucester. A hurricane, unknown in Europe, arising in that rapid river, carried the boats down some miles from the crossing to Gloucester, and directly under the lines of the enemy, and into the van of the fleet; the boats crew abandoned their oars at once, and trusted to their skill in swimming, when this brave youth, undismayed, and determined rather to perish than to become prisoner, by a manoeuvre and labour scarce conceivable, carried these boats into creeks. Here laying the men on their faces under the rocks, not breathing to be heard by the enemy, whose lines extended over them, they weathered the storm, and saving just the darkness of the night, got off with their boats, which only by landmen were rowed back to York-Town, the crossing to Gloucester being impracticable by the eddy; for which conduct the

Captain received the utmost testimony of applause and esteem from Lord Cornwallis.[51]

Regardless of Asgill's role in the action, the storm's impact marked the end for Cornwallis's encircled army. Cornwallis described the dilemma he faced on the morning of 20 October:

> In this situation, with my little force divided, the enemy's batteries opened at day break: The passage between this place and Gloucester was much exposed, but the boats having now returned, they were ordered to bring back the troops that had passed during the night, and they joined in the forenoon without much loss. ... Our numbers had been diminished by the enemy's fire, but particularly by sickness; and the strength and spirits of those in the works were much exhausted by the fatigue of constant watching and unremitting duty. Under all these circumstances, I thought it would have been wanton and inhuman to the last degree to sacrifice the lives of this small body of gallant soldiers, who had ever behaved with so much fidelity and courage.[52]

On that day Cornwallis asked his opponents to negotiate terms of capitulation. On 19 October a surrender ceremony saw British, German and Loyalist soldiers lay down their arms before the combined American and French armies. Claiming illness, the British general sent his deputy, Foot Guards officer Brig. Gen. Charles O'Hara, to surrender on his behalf. O'Hara approached the commander of the French troops to surrender his sword, the Frenchman deferred to Washington, who in turn gave the nod to his own deputy, Gen. Benjamin Lincoln, who accepted it.

Asgill had spent exactly four months at war in America, during which time he may have shown great strength of character and honour, from landing on 19 June until walking with his troops to surrender on 19 October. This capitulation had far-reaching consequences, forcing Britain into peace negotiations. America had gained her independence and Britain had lost valuable colonies in North America. Asgill's American Westminster school-friends, now serving in General Washington's army, had won their battle. But that friendship survived to live another day.

Officers, treated as gentlemen by their captors, were offered paroles – as long as they did not participate in any way in the conflict, they were free to return to Europe, their word of honour being sufficient security. But the prisoners of war needed supervision, so a lottery determined which officers of each rank would remain in America to see to the welfare and discipline of their men, retaining a military structure for the captive army.

    Captain Asgill was one of the officers of the Brigade of Guards chosen to stay with the captives. Lieutenant Colonel Lake drew the lot as one of the three senior British officers to remain. Lake negotiated a change due to having private affairs to settle, and Major James Gordon of the 80th Regiment of Foot agreed to take his place.[53] Gordon could not have anticipated that a much more important "lottery" was on the horizon, in which he would become heavily involved. Who knows how Lake would have handled it, but with the wisdom of hindsight, Asgill won the lottery that day in more ways than one, for Gordon ultimately was instrumental in saving Asgill from death.

CHAPTER FIVE
Captain William McCurdy

After capitulation, Asgill became a prisoner-of-war. The captive army was marched to York, Pennsylvania, well away from front lines. The soldiers were held in a stockade, while officers like Asgill were granted local paroles and were allowed to keep their soldier servants with them. These officer-prisoners were treated as men of honour, but their freedom of movement was limited; they could not bear arms and could not, yet, go home. Asgill's duties at this time were to look after the welfare of his men, to ensure they were provided with food and shelter until they could either be exchanged or repatriated. Although Yorktown was the last major military action of the war in America, no peace had yet been agreed upon. For the British and German prisoners, there was no immediate prospect of freedom.

While on parole in Pennsylvania he had the misfortune of coming across an American officer; a 52-year-old Captain William McCurdy of the 1st Pennsylvania Regiment. McCurdy was in the area on recruiting service, away from his regiment and in a position to interact with prisoners on parole. To say that McCurdy and Asgill had "words" would be understating the situation – no record of their specific altercation or altercations has been found, but their "broils and quarrels"[54] were severe enough that Asgill made a formal complaint to the commanding officer of the British prisoners, Major James Gordon. Gordon, responsible for the welfare of those under him, considered the matter serious enough to bring it to the attention of Brigadier General Moses Hazen, the American officer charged with overseeing the prisoners. Asgill's complaint had sufficient merit that Hazen put McCurdy under arrest. The matter now required a court martial, initiating a confusing, and to many of those involved, confounding, series of events.

To conduct a court martial required considerable time and effort. An officer with the role of Deputy Judge Advocate (deputy because he reported to the army's overall Judge Advocate General) spoke to both the plaintiff and the defendant, gathered evidence for both the prosecution and defence, notified witnesses, and ensured that the trial was conducted according to established procedures. A board of officers – thirteen for a general court martial, fewer for garrison, brigade or regimental courts – convened to hear evidence, determine guilt or innocence, and recommend a sentence. The senior-most of those officers was the court's President, in overall charge of its conduct in accordance with the procedures enforced by the Deputy Judge Advocate. The Deputy Judge Advocate saw to it that the proceedings were recorded, and sent the proceedings to the Commander-in-Chief, who either approved the sentence or called aspects of the proceedings into question.

General Hazen saw an opportunity to simplify this process by having the case heard by a court that was already sitting in Lancaster, about 25 miles from York and a town where, like York, British prisoners of war were held. It seemed like a convenient choice. For reasons that aren't clear, McCurdy refused to have his case heard by this court. This was a defendant's privilege – he could object to individual members of the court, or to the composition of the court in its entirety, if he had grounds to suggest that the trial would be unfair. But he may have simply hoped to delay the trial, knowing the challenges of convening a new court just for him.

Another court convened in Carlisle, Pennsylvania, about 40 miles northwest of York on 18 March 1782. The President was Colonel Richard Butler of the 5th Pennsylvania Regiment, an officer McCurdy probably knew well. The distance from York made it cumbersome for the plaintiff – Asgill – and witnesses on his behalf, probably fellow prisoners, to appear and testify. These British prisoners-of-war, although on parole, did not have their own freedom to travel such a distance, placing a burden on Colonel Hazen to facilitate their journey and make arrangements for sufficient custody during the trial. So Asgill did not "totally (& rather contemptuously) neglect" to attend the court, as Butler insinuated.

The court agreed to hear McCurdy's case first, but the prosecution did not have any evidence to present – Asgill and any witnesses on his behalf were not in Carlisle. According to Butler, "want of evidence Induced the court to defer the trial from time to time in order that the british officer might Appear with his Evidence to prosecute which was totally (& rather contemptuously) neglected till all the other prisoners were tried." On 16 May the court was again ready to hear the case, but still there was no prosecutorial evidence. "The court having no other business to justify further protraction of trial," wrote Butler, "were Obligd to proceed to A Conclusion."[55]

Butler was clearly not satisfied with the way the case was handled, and laid the blame on Hazen. Writing to General Washington about the findings, he said, "The case of Captn McCurdy I consider to be of a very peculiar & delicate nature & Sincerily wish the Affair had been conducted in A different maner than that which Appeared to the court improper & officious on the part of general Hazen." This may refer to Hazen's failure to get the plaintiff or his witnesses to the trial. Butler also took issue with "the right of the prisoner in his choice of trial" which he considered an attack on "the dignity & authority of General Courtsmartial." Something in Hazen's letters concerning the change of venue from Lancaster to Carlisle apparently came across as insulting to the officers of one or both courts, for Butler complained of "their Veracity as officers Obliquely Struck at by insinuations which had no grounds."[56]

Washington's response to Butler on 10 June showed that the matter was not straightforward: "The Trial of Capt. McCurdy I have not had Time fully to consider—at first reading however it appears singular in all its parts—I shall take time to give it a more attentive perusal, & inform you of my Opinion."[57] On 23 June Washington announced in general orders that he disapproved of the proceedings of the court, because of "great irregularity in the mode of bringing Captain McCurdy of the Pennsylvania Line before the Court Martial." It was clear that Washington believed the charges against McCurdy had merit, for he ordered a new trial to be held in Philadelphia "as soon as circumstances will admit."[58]

Washington tasked Benjamin Lincoln, the Secretary of War, with appointing officers to form a new court, partly because, as he wrote to Lincoln, "I do not consider Brig. General Hazen as commanding through out the State of Pennsylvania & consequently capable of Ordering Officers from the various posts of the State to attend at whatever place he may think proper," and partly because "I have great reason to apprehend a spirit of faction & other ill consequences will be produced by the clashing sentiments of the different Officers concerned in the dispute, to the great detriment of service, unless measures are taken to prevent it." McCurdy's objections to the first court at Lancaster, and Hazen's handling of the transfer of the case to the court at Carlisle, had ruffled many feathers.[59]

Hazen, clearly offended by the apparent slight to his abilities as an officer, wrote a long letter to Washington on 10 July. He explained that he had asked the Secretary at War for advice or orders on how to handle the case, and then was "favored with your Excellencys orders of the 23rd of June" that disapproved of the sentence of the Carlisle court and directed the trial to be held in Philadelphia. He assured Washington that he "never had an Idea of Commanding anything more in this State, than my own Regiment, and the prisoners of War." That said, he asserted,

> If Officers of other Corps come within the limits of my Command, and plunge themselves into broils and quarrels, with those under my immediate protection, Your Excellency will, I am pursuaded, agree with me, that in such cases, duty would require my intoposition, more especially, when justice is not only due to individuals; but the honor of the American Army concern'd.[60]

He pointed out that he had never seen or even heard of McCurdy until Asgill's complaint was brought to him, and that he did not have any "unjust prejudices against him." He did, on the other hand, think that Colonel Butler's conduct was "odd," and he "was not well satisfyed with what appear'd a want of civility, at least toward me." Hazen told Washington that it was Butler, not he, who ordered that Asgill and his witnesses attend the trial in Carlisle, something that they, as prisoners of war, could not do.[61]

Butler sent his own letter to Washington the next day, telling Washington that "the court felt themselves hurt by the Strictures expressd in the order" that disapproved of their proceedings. He also defended his own actions by saying that he "did not see the way of the court clear in taking up the tryal of Captn McCurdy," but that it was the court's decision, not his own, to take the case.[62]

By this time, though, there was yet another obstacle to holding a trial: Charles Asgill, the complainant, had been taken into confinement late the previous month for reasons unrelated to the McCurdy trial that will be discussed in the next chapter. Washington was well aware of this, telling Lincoln that if Asgill's circumstances "finally debar him from carrying on the prosecution; I should presume the charges will Lie of course, & Capt. McCurdy must be released from his arrest—this will probably be ascertained in the course of a short time." He closed by apologizing for troubling Lincoln with the matter, even though he trusted Lincoln to be "always ready to incur any personal trouble which will contribute to the public good."[63] Lincoln gave a knowing response that he would convene a trial for McCurdy "should it be necessary hereafter."[64]

The matter was not resolved "in the course of a short time." It remained in limbo for four months because a proper trial could not be conducted without Asgill's presence. As will be seen, the affair was eclipsed by more serious events, but McCurdy nonetheless remained under arrest awaiting trial. Finally on 21 November 1782 Washington acknowledged "the long and disagreable restraint he [McCurdy] has laboured under and the many difficulties which would attend the assembling a General Courtmartial for his trial," and ordered him released from arrest to return to duty. To make sure that McCurdy was free from condemnation as well as from confinement, Washington added that he thought "proper to declare that Captain William McCurdy is as free of any imputation of misconduct as if his arrest had not taken place, or had been acquitted on the fullest investigation of his case by as General Courtmartial."[65]

It is not known what Asgill may have thought about this. Or, indeed, Hazen. Although, as will become clear, Asgill soon came to realise that he had great respect for Brigadier General Moses Hazen.

CHAPTER SIX

Condemned to be Hanged

The American War of Independence is often presented as a straightforward conflict between two nations, a series of battles between well-defined armies. Given his limited time in the country, Asgill himself may have seen it that way, at least initially. The reality in many regions was quite different. Differences between those who saw the American colonies as subordinate to the British government and those who saw American independence as an escape from tyranny tore apart communities and even families. The inherent polarisation of war put in danger those who wished to remain neutral. Settlements near or between front lines were dangerous places, sometimes for years on end, as soldiers raided and foraged, and opportunists took advantage of chaotic conditions. The relative lawlessness of war allowed old disputes to be settled with violence.

The coastal regions of New Jersey were among the places where violence was frequent and loyalties were often fluid. Even after the British surrender at Yorktown, in early 1782 small-scale but fierce warfare still raged there, often between small parties of militia and refugees with no sanction or even awareness of higher military authorities. In late March, Philip White, a New Jersey Loyalist who had fled to New York and taken up arms in support of the British Army, was captured by American soldiers in New Jersey. Accounts vary as to whether he was attempting to visit family members or was participating in a raid. Rather than treat him as a prisoner of war, his Patriot captors brutally murdered him.

Enraged Loyalists vowed revenge. An American officer, New Jersey militia Captain Joshua Huddy, was soon implicated in White's murder. Huddy had a reputation as a fierce marauder who had murdered other Loyalists; even though he had been captured in 1782 and was apparently in custody in New York when White was killed, he became the target of vengeance. Arrangements were made to exchange Huddy, but when he was taken from New York to the New Jersey shore on 12 April to be released, his captors instead left him hanging from a tree with a notice reading:

> We the refugees, having with grief long beheld the cruel murders of our brethren, and finding nothing but such measures daily carrying into execution — we, therefore, determine not to suffer without taking vengeance for the numerous cruelties; and thus

begin, and, I say, may those lose their liberty who do not follow on, and have made use of Captain Huddy as the first object to present to your view; and further determine to hang man for man while there is a refugee existing. Up goes Huddy for Philip White.

It was time for General Washington to intervene, perhaps because the major battles had finally been won, and perhaps because an eventual peace necessitated ending these tit-for-tat encounters. But Washington made one of the most controversial moves of his career, arguably his biggest military-diplomatic misstep. Rather than demand an outright stop to the violence, he wrote that "this Instance of Barbarity, in my Opinion, calls loudly for Retaliation."[66]

On 21 April, Washington sent a letter to the British commander in chief, General Sir Henry Clinton, demanding that the perpetrator of Huddy's murder be handed over to American authorities. He included copies of depositions from several people who had, or claimed to have, first-hand knowledge of the affair, calling it "the most wanton, unprecedented, & inhuman Murder that ever disgraced the arms of a civilized People," even though it was the most recent of many such crimes by people on both sides. If the perpetrator was below Huddy's rank of captain, Washington demanded "so many of the perpetrators, as will, according to the Tariff of Exchange be an Equivalent"; that is, following protocols that placed a value on ranks for the purposes of prisoner exchanges, Washington wanted either a captain or more than one man of inferior rank.[67] There was a darker side to Washington's intentions. He informed a member of the Continental Congress that if the British commander refused, "I should be driven to an Act of Retaliation, a British officer of equal Rank, must atone for the Death of the unfortunate Huddy."[68]

Clinton did not acquiesce to Washington's demand. He explained in a letter on 25 April that he became aware of Huddy's murder well before receiving Washington's letter, and was determined to bring the perpetrators to trial. He admonished Washington's use of "very improper Language" which "You could not but be sensible was totally unnecessary." He made it clear to Washington that he would not respond to an unlawful, unsanctioned act by committing another: "To Sacrifice Innocence under the Notion of preventing Guilt, in Place of suppressing, would be adopting Barbarity and raising it to the greatest height."[69]

The Charles Asgill Affair

On 3 May 1782, Washington sent instructions to Brigadier General Hazen, the officer responsible for British prisoners of war in Pennsylvania:

> The Enemy, persisting in that barbarous line of Conduct they have pursued during the course of this war, have lately most inhumanly Executed Capt. Joshua Huddy of the Jersey State Troops taken Prisoner by them at a Post on Toms River—and in consequence, I have written to the British Commander in Chief, that unless the perpetrators of that horrid deed were delivered up, I should be under the disagreeable necessity of Retaliating, as the only means left to put a stop to such inhuman proceedings.
>
> You will therefore immediately on recet of this designate by Lot for the above purpose—a British Captain who is an unconditional Prisoner, if such a one is in our possession—if not—a Lieutenant under the same circumstances, from among the Prisoners at any of the Posts either in Pennsylvania or Maryland.
>
> So soon as you have fixed on the Person—you will send him under a Safe Guard to Philadelphia, where the Minister of War will order a proper Guard to receive & conduct him to the place of his Destination.
>
> In your information respecting the Officers, who are Prisoners in our possession, I have orderd the Commisy of Prisoners to furnish you with a List of them—it will be forwarded with this.
>
> I need not mention to you that every possible tenderness, that is consistent with the security of him, should be shewn to the person whose unfortunate lot it may be to suffer.[70]

Washington specified "unconditional prisoners," meaning prisoners who were captured but not held under the terms of any treaty or convention. Hazen was unable to comply, because there were no such officers under his jurisdiction. His prisoners had surrendered at Yorktown under a treaty that included a specific article:

> Article XIV. No article of capitulation to be infringed on pretence of reprisals; and if there be any doubtful expressions in it, they are to be interpreted according to the common meaning and acceptation of the words.[71]

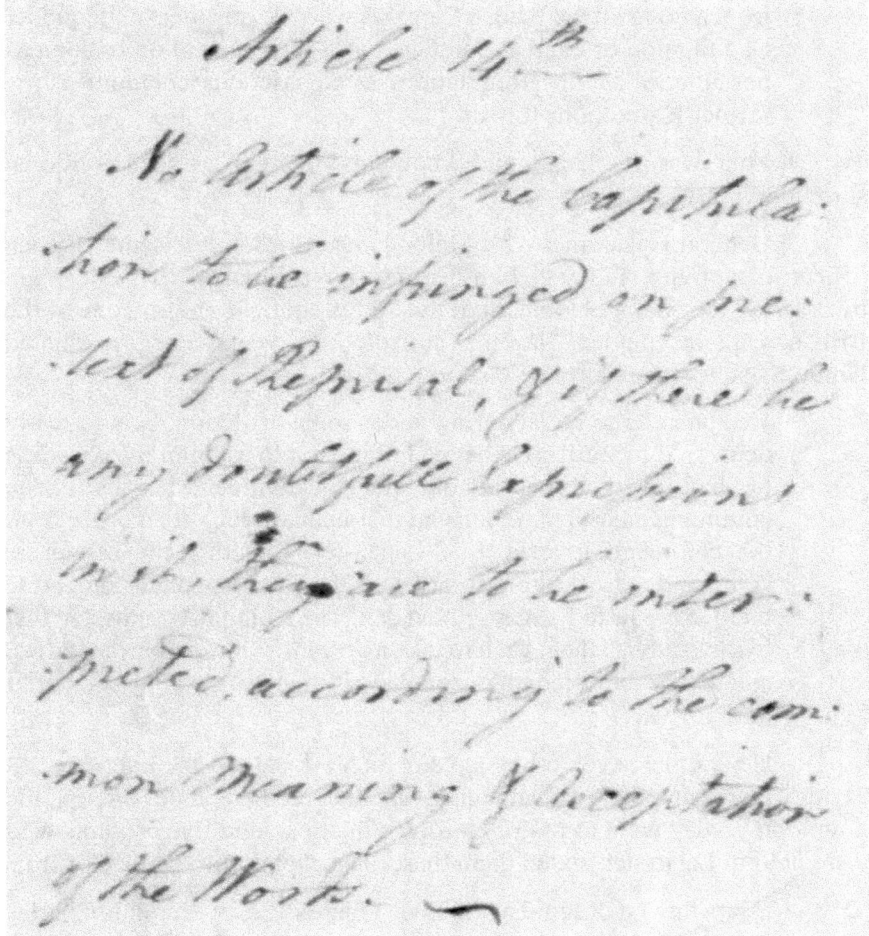

Article XIV of the Articles of Capitulation of the British at Yorktown, 19 October 1781. George Washington Papers, Library of Congress

Unwilling to change course, and getting impatient at the lack of action, Washington sent another order to Hazen on 18 May:

> It was much my Wish to have taken for the purpose of Retaliation, an Officer who was an unconditional prisoner of War—I am just informed by the Secty at War, that no one of that Description is in our power—I am therefore under the disagreeable necessity to Direct, that you imediately select, in the Manner before presented,

from among all the British Captains who are prisoners either under Capitulation or Convention One, who is to be sent on as soon as possible, under the Regulations & Restrictions contained in my former Instructions to you.[72]

This was a blatant and knowing violation of the Yorktown surrender treaty.

General Hazen met with Major James Gordon, the senior British officer in captivity, and gave him instructions to choose a British captain by drawing lots. Samuel Graham of the 76th Regiment of Foot, one of the British captains under Hazen's jurisdiction, years later recounted Gordon's subsequent surprise visit to him:

He appeared to be labouring under some affliction, being greatly depressed in spirits. He begged of me not to ask him the cause, as he had pledged his honour not to divulge what had been communicated to him, but said that he had brought an order from the commanding officer at Lancaster, directing the officer in command at Little York to order all the British captains on parole there to repair to Lancaster next day. The major also requested that I would advise them each to take a servant, with spare necessaries, and that he expected to see them at his quarters next day soon after their arrival.[73]

He and the seven other captains in York set out for Lancaster on 25 May, "crossing the Susquehannah, and arrived there about three in the afternoon." They went to Major Gordon's quarters, and five captains who were held in Lancaster joined them there. The captains assembled, were:

From the 1st Regiment of Foot Guards: Lt. & Captain Charles Asgill, Lt. & Captain Hon. George Ludlow, Lt. & Captain James Perrin

From the 2nd Regiment of Foot Guards: Lt. & Captain George Eld, Lt. & Captain Henry Greville

From the 23rd Regiment of Foot (Royal Welch Fusiliers): Captain Thomas de Saumarez

From the 76th Regiment of Foot, Captain David Barclay: Captain Samuel Graham

From the 17th Regiment of Foot: Captain Lawford Miles

From the 26th Regiment of Foot: Captain Bulstrode Whitlocke

From the 33rd Regiment of Foot: Captain James Ingram

From the 80th Regiment of Foot: Captain Alexander Arbuthnot, Captain William Hawthorn (also known as John Hathorn)[74]

Asgill was the youngest of them all, having just turned 20. He and the other Foot Guards officers were there only by a quirk of army ranks. In their own regiments, they were lieutenants, but Foot Guards officers were accorded seniority over officers of the regular army, so lieutenants in the Guards also ranked as captains in the army. What was otherwise a benefit was now a curse for Asgill, Ludlow, Perrin, Eld and Greville.

Major Gordon paced the room, not knowing how to put the news across to his rapt audience. They were all aware of the correspondence between Washington and General Clinton about Huddy's murder, for it had been published in newspapers that the paroled prisoners of war had ready access to. And yet they did not feel in danger, knowing that they were held under the terms of a treaty that protected them from retaliatory punishments. They were astonished when, as Captain Graham wrote, "The major addressed us in a most feeling manner, acquainting us that orders had arrived to send on one of us as a subject of retaliation for the murder of a Captain Huddy."[75]

Asgill's friend Henry Greville wrote to his mother that they all owed a debt of gratitude to Gordon for the manner in which he handled this terrible situation.[76] Gordon addressed the matter as directly and gracefully as he could:

> Gentlemen, you will scarcely believe that in face of the capitulation, and in defiance of the strong remonstrances which I felt it my duty to make, both to the American and French authorities, one of you is doomed to suffer. I have told Washington that he will be answerable for this foul deed to all posterity: but I might as well reason with the air! I wish to God they would take me in your place; for I am an old worn-out trunk of a tree, and have neither wife nor mother to weep for me. But even to that they will not consent; so all that I can undertake to do is, to accompany the unfortunate individual, whoever he may be, to the place of his martyrdom, and to give him every consolation and support while life remains, and obey his wishes after it is taken away.[77]

The Charles Asgill Affair

Gordon, approximately 47 years old, did not have a wife, but he certainly had a loving family consisting of a mother, a brother and four beautiful sisters. He was telling an honourable lie to support the young officers from whom he must choose one to die. One of the officers present remembered, "The Major, albeit not given to the melting mood, could not here restrain his tears; and there was not a soul among us who did not feel a thousand times more for him than for ourselves."[78]

Foreboding though the news was, the officers "spent a tranquil and even lively evening at his table." They were as concerned for Gordon as for themselves, Graham writing that he slept well "after having vainly striven to cheat my excellent friend into a forgetfulness of the care and anxiety which preyed upon his mind."[79]

The following morning "a like, perhaps a more exuberant, spirit pervaded us" when they rose to face whatever fate might have in store for them at the Black Bear Tavern in Lancaster. The group assembled there at nine o'clock for the drawing of lots.

Hazen began by reading his orders from Washington, his voice faltering as he began. After the letters were read, he asked the officers to decide among themselves "on whom this sad fate should be fixed." The captive captains refused, unwilling to have a part in an affair that was in violation of the surrender treaty. Major Gordon went so far as to point out that many other captains had become prisoners at Yorktown, and if time was allowed they would certainly come from England to stake their lives alongside those present. All Hazen was able to say was that he deeply regretted the need for him to obey the orders issued by his commander-in-chief, but that he would do all within his power to ease the passage for the selected victim.[80]

The American officers left the room to prepare the lots. The British officers waited in near silence. After about ten minutes Hazen returned with his aide-de-camp and the commissary of prisoners, each carrying a hat and accompanied by a young drummer, or perhaps two (accounts on this matter vary between one to three). Also with them was an armed dragoon officer. In one hat were 13 slips of paper with the names of the captains, in the other 12 blank slips and one marked "Unfortunate." One by one, a drummer drew a name and the other drew a fate, until the eleventh draw revealed Charles Asgill to be the Unfortunate.

Hazen allowed the British officers to stay in Lancaster on the night of 27 May, which gave Gordon time to finish writing many letters to influential people, and for many of the captains to write home to tell their families of these events, just as Henry Greville did before he left to return to York. Asgill chose not to do so, and left the matter hanging until he thought his end was near.

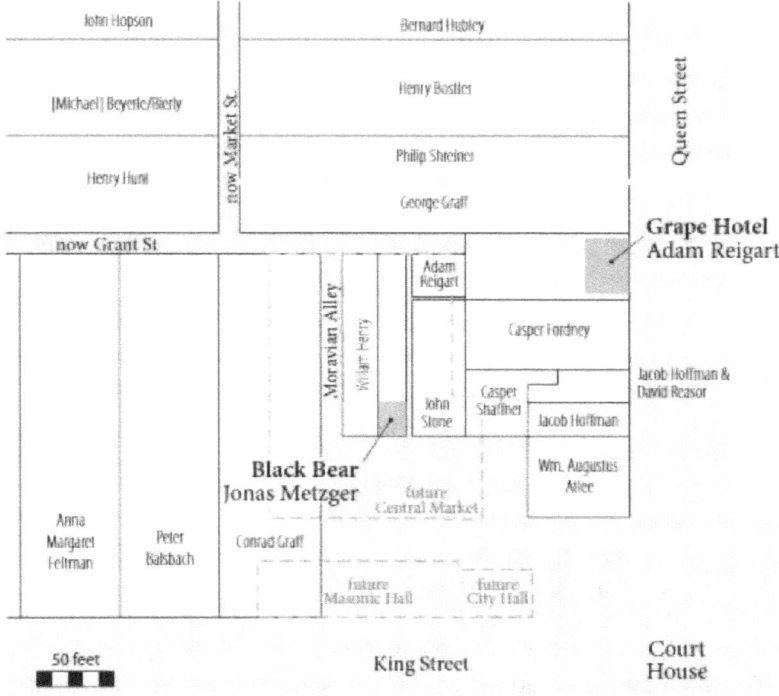

Market Square, Lancaster, Pennsylvania, in 1782, showing location of the Black Bear tavern. Dashed lines indicate present-day buildings. Courtesy of LancasterHistory, Lancaster, PA.

Hazen was a good and decent man of honour, so he kept his promise to them. He allowed Captain Ludlow to leave Lancaster ahead of the others in order for him to deliver the letters Major Gordon wrote to officials on the British, American and French side, all pleading with them to find a way to save the young man's life. All of these letters had been written the night before with only a blank left to be filled with the name of the "Unfortunate" officer. One was, of course, written to Washington:[81]

Lancaster 27th May 1782

Sir

It is with astonishment I read a Letter from your Excellency, dated 18th May, directed to Brigadier General Hazen, Commanding at this Post, ordering him to send a British Captain, taken at York–town by Capitulation, with My Lord Cornwallis, Prisoner to Philadelphia, where 'tis said he is to suffer an ignominious Death, in the room of Capt'n Huddy an American Officer, who was murder'd by a Lawless Banditti, calling themselves Refugees.

As this is, in Direct Violation of the Articles of Capitulation of the Garrison's of York and Glocester, which your Excellency signs first, as Commander in Chief of the Combin'd Forces of America and France, the 14th Article of which, expressly says "That no Articles of the Capitulation are to be infring'd on Pretext of Reprisals, and if there be any doubtfull Expressions in it, they are to be interpreted according to the common meaning and acceptation of the Words." I therefore, in the Name of my Royal Master George the Third, King of Great Britain, Demand of Your Excellency, that you will order Captian Asgill of the Brigade of Guards to be discharged from his Confinement, and admitted to the same Parole, as the Other Officers of my Lord Cornwallis's Army.

I have the Honor to be Your Excellency's Most Obed't & most Hum'le Serv't

James Gordon, Major 80 Regt & Field Officer with the British Prison.

Time was needed to bring this insanity to an end, and Hazen bought some time. He also wrote letters expressing his deep regret at this turn of events. In addition, he provided Gordon with names of those in Philadelphia who might fight in Asgill's corner.[82]

Hazen's letter to Washington was long, ending thus:[83]

Since I wrote the above Majr Gordon has furnished me with an Original Letter of which the inclosed is a Copy, by which you will see we have a Subaltern Officer and unconditional Prisoner of War at Winchester Barracks. I have also just received Information that Lieut. Turner, of the 3rd Brigade of Genl Skinner's New-Jersey Volunteers is in York Goal—but as those Informations did not come to Hand before the Lots were drawn, and my Letters wrote to your Excellency and the Minister of War on the Subject, and as I judge no Inconveniency can possibly arise to us by sending on Capt. Asgill, to Philadelphia, which will naturally tend to keep up the Hue and Cry, and of course foment the present Dissentions amongst our Enemies, I have sent him under guard as directed. Those Officers above-mentioned are not only of the Description which your Excellency wishes, and at first ordered, but in another Point of View are proper Subjects for Example, been Traitors to America, and having taken refuge with the Enemy, and by us in Arms. It have fallen to my Lot to superintend this melancholy disagreeable Duty, I must confess I have been most sensible affected with it, and most sincerely wish that the Information here given may operate in favour of Youth, Innocence, and Honour.

The plea was not only a graceful show of humanity by Hazen, but also pointed out the poor manner in which the selection of a victim was handled. Washington had originally asked for an unconditional prisoner, but then was told by the Secretary of War that none were available. Only after Asgill's lot was drawn did Hazen learn of two other officers – subalterns, that is, below the rank of captain – in captivity who, because they were captured in arms rather surrendered under a formal treaty, were better suited to Washington's wishes.

General Hazen nonetheless followed orders to send the condemned man to the designated place of execution, Chatham, New Jersey, only a few miles inland from the coast and from a British garrison on Staten Island. He acceded to Major Gordon's requests by laying out a route of many short stages, thereby taking more time than might otherwise be necessary, and allowing Gordon to accompany Asgill. Hazen too accompanied them for the first few miles, and before departing ordered the dragoons escorting the prisoners to adhere to all commands from Gordon, save any that might compromise the prisoner's security, and they must prevent his escape. Hazen was Asgill's first "Wayside Samaritan."[84]

The 70-mile journey to Philadelphia took a couple of days. It would have been uneventful, "was I not," as Asgill later wrote, "publickly exhibited & unnecessarily paraded with an Escort of Dragoons in to Philadelphia & the Towns in my way to the Jerseys."[85] These public humiliations, even though Asgill had done nothing wrong, were a prelude to hardships to come – the Huddy incident was big news in America, and vengeance was on many minds.

Upon arrival in Philadelphia on 30 May, Gordon found lodgings for the two of them. Asgill applied himself to writing a letter to George Washington, pointing out that he was a protected officer and, as such, his life should not be in danger for reasons of retaliation. Under his signature he wrote, "Lieutt 1st Regt Guards," subtly asserting that he was not even a captain, at least not in his regiment.[86] Major Gordon, meanwhile, placed a sentry at the door of Asgill's room with orders to let no one molest him. He then "lost not a moment in finding the French ambassador, urging him in strong terms to interfere. He also found out some members of Congress, and applied to them; in short, he tried every possible means which he could think of or devise."[87]

There was no immediate or apparent result from any of these entreaties, and Gordon eventually found his options exhausted. He did have one experience which may not have amused him at the time, but which he was later able to recount with a laugh:

> Utterly dejected, he returned home, and had thrown himself upon his bed, when the sound of footsteps approaching Asgill's chamber roused him; he ran out, and beheld a tall gaunt figure, arrayed in black, with an expression of singular austerity in his countenance, advancing with measured tread towards the door.
>
> "Who are you —— what do you want?" were the brief questions.
>
> "Sir," replied the figure with extreme solemnity, "I am chaplain to the Congress of the United States, and I am come to give a word of advice to the young man who is about to suffer for the death of our good countryman, Captain Huddy."
>
> The Major was a religious man, in the best sense of that term, and entertained unfeigned respect for the clergy; but his temper was at the moment rendered irritable by his recent repulses, and the manner in which the divine spoke of the approaching murder of his friend threw him entirely off his guard.

"I tell you what," cried he, springing forward, "if you do not immediately remove yourself from this house, I will show you the shortest way into the street, even if it should be from the window."

The divine looked aghast, and retreated as the other drew on, till he gained, without being aware of it, the top of the staircase, when suddenly his back step failed him, and he rolled from the top to the bottom. No further mischief followed, however, except the loss of his hat and wig, both of which fell off in the tumble, for Mass John was not long prostrate: he rose immediately, and apprehending he knew not what further violence, grasped the wig, clapped it wrong end foremost upon his head, and holding his hat in his hand, ran with the speed of a lamp lighter down the street.[88]

CHAPTER SEVEN

Under Close Arrest

After a couple of days in Philadelphia they departed, heading northeast towards Chatham, New Jersey. It was a long journey to Chatham from Lancaster, around 150 miles, and the two officers spent the time talking and wondering how things would turn out. In Gordon, Asgill knew he had the very best counsellor; one who guided and advised him in all aspects of what would follow. He couldn't possibly have been in better hands and his love and gratitude to this man stayed with him for the rest of his life.

When they arrived in Chatham on approximately 7 June, if not later, they were met by Colonel Elias Dayton, commander of the 2nd New Jersey Regiment in the Continental Army. He was courteous and hospitable to them both and, it seemed, they were encountering yet another Wayside Samaritan; a man who until less than a year ago had been their enemy, and them his. Indeed, he housed them in his own personal quarters and looked after them well, while at the same time realising the reason for them being under his command. He permitted them to be on parole, free to come and go to some extent. They rode their horses for exercise and returned to be under his orders at the end of each day.

This didn't last long though. News of this arrangement reached Washington's ears and, on 11 June 1782, just days after their arrival, he wrote to Dayton:

> I am informed that Capt. Asgill is at Chatham, without Guard, & under no constraint — This if true is certainly wrong — I wish to have the young Gentleman treated with all the Tenderness possible, consistent with his present Situation — But untill his Fate is determined, he must be considered as a close prisoner & be kept in the greatest Security[89]

This was the turning point of what became known as the Asgill Affair. Washington ordered Dayton to send Asgill to a military camp to be "kept close prisoner, in perfect Security 'till further Orders."[90]

Dayton responded on 17 June that he would immediately send Asgill to "the Jersey Huts," the cantonment of the New Jersey brigade in the area. But upon seeing Asgill the next day he found him "in such a situation that humanity would have shuddered at the idea of his removal he has been in a fever for some time past and the agitation of his mind upon the apprehension of less agreeable quarters and perhaps more indelicate treatment have increased it to a very considerable degree." He explained that he would obey his orders, but only when Asgill was well again.[91]

Colonel Elias Dayton's House in Chatham, New Jersey, with kind permission of Summit Historical Society. Wikimedia Commons.

There was another factor working in Asgill's favour, at least in terms of buying him time. The British commander in chief, General Clinton, had made good on his promise to find the perpetrator of Huddy's murder and bring him to trial. Captain Richard Lippencott of a corps called the Associated Loyalists was arrested, and on 14 June his trial by court martial in New York began. He was charged with "The Murder of Joshua Huddy a Prisoner of War to the Associated Loyalists, by hanging or causing him to be hanged by the neck until he was dead. He (Lippincott) having received the said Huddy into his custody for the purpose of exchanging him for another prisoner of war."[92]

In addition, there was a change in the British command. General Clinton resigned his post as commander in chief of British forces in North America, and was replaced by General Sir Guy Carleton. Carleton had long been the colonial governor of Quebec and was highly experienced in diplomatic as well as military matters. Asgill's close friend and fellow captain, George Ludlow, had been permitted to go to New York to deliver letters written on Asgill's behalf and to make the case for negotiation on the matter. Carleton, rather than Clinton, was now the one to handle the case. In his 11 June letter to Dayton, Washington mentioned this development, but informed Dayton that he had "heard noting [sic] yet from N. York in Consequence of this Application." But Washington closed by restating his firm commitment to retaliation: "His Fate will be suspended 'till I can be informed the Decision of Sir Guy but I am impatient least this should be unreasonably delayed — the Enemy ought to have learnt before this, that my Resolutions are not to be trifled with."[93]

When Asgill was well enough, Dayton moved him to new quarters. Up until then Asgill had met with kindness and although Washington maintained that he wished for him to be treated "with tenderness," this turn of events ensured that everything changed very much for the worse. It is not known why Asgill was not sent to the Jersey Huts, as Washington had ordered. Perhaps there was no suitable place among the soldiers' cabins for holding a "close prisoner" in the "greatest security." Instead he was sent to a public house close to Dayton's home, in what is now known as Summit, New Jersey, a district within Chatham.[94] He was under close arrest and the whole tenor of his captivity was now under the orbit of the landlord of the of the public house.

This public house was almost certainly Timothy Day's Tavern, the closest one to Dayton's house.[95] The tavern was a hotbed of revolutionary ardour where soldiers congregated; after a 1780 battle in New Jersey, wounded soldiers were brought there.[96] A liberty pole stood in front of the building.[97] Asgill's presence gave the revolutionaries the ideal opportunity to make his life as miserable as they possibly could. It certainly didn't help matters that the trial of Captain Lippincott concluded on 24 June, with a verdict that inflamed Americans:

> it appearing that (altho' Joshua Huddy was executed without proper Authority) what the Prisoner did in the matter was not the effect of Malice or Ill-will, but proceeded from a Conviction that it was his Duty to obey the Orders of the Board of Directors of Associated Loyalists, and his not doubting their having a full Authority to give such orders, the Court is of Opinion that he, the Prisoner, Captain Richard Lippincott, is Not Guilty of the murder laid to his charge, and doth therefore Acquit him.[98]

Under Close Arrest

The Associated Loyalists, responsible for Huddy's death, was an organisation headed by William Franklin, Benjamin Franklin's acknowledged illegitimate son and the last royal governor of New Jersey. William had previously signed a warrant to hang Huddy in retaliation for the murder of the Loyalist, Philip White, which was the caveat which saved Lippincott from being handed over to Washington by the British.[99]

Timothy Day's Tavern. Drawing by Mary Keim Tietze, 1967, based on archive records. This was the location of Asgill's imprisonment in 1782. With kind permission of the Chatham Historical Society.

But the clientèle of the tavern had Huddy's murder to avenge, and made the most of opportunities to bribe their way into Asgill's small cell. A little liquor for the guards, or largess for the proprietor, was all it took to gain access to the helpless prisoner at any hour of day or night. They came into his room and taunted and teased him, and often this turned violent. One of these unwanted "visitors," when Asgill refused to give him the answers to his questions that he wanted, got so angry that he beat Asgill violently; Asgill wrote that he nearly died from his injuries.[100]

He was also not fed in a way to sustain life, having only bread and water delivered to him. Sleep was a rarity since the torment mainly occurred during the night-time. As always, Major Gordon came to his aid, somehow obtaining a large sum of money from the British quartermaster general to finance ways for easing Asgill's suffering. A year later Gordon explained:

> I not only thought it my Duty to accompany him during his confinement but likewise to provide such a Sum of Money as might be sufficient to defray the Expenses of that unfortunate affair and if possible contribute to his Release or Enlargement.[101]

This letter may well have been the last written by James Gordon, since he died five days later; thinking of Asgill to the last. So, as soon as the landlord was aware of this money, he provided edible food at the most exorbitant price.[102]

The days, weeks and months went on like this. Asgill awaited the end to his life, daily, for six months, rising each morning to wonder if it would be his last. But Major Gordon was plotting a way out. Not only did he write many letters, especially to Carleton, seeking a solution to the predicament, but laid a plan about which he naturally told Asgill very little.

> With the help of certain American ladies whose friendship he had reason to know and to trust, Gordon had laid a well-ensured plan for the spiriting away of Asgill, should the order for his execution come through. This plan, in its entirety, could not be confided to Asgill, since, had he guessed its full nature, he would certainly have refused to stir. For it involved, for Gordon, the final sacrifice. The boy once safely off beyond recall or the possibility of return, Gordon, avowing his own authorship of the escape, would have stepped into the place of the condemned. Such, in essence, was his purpose.[103]

Gordon even took the bold step of asking for help from Huddy's widow and family. She certainly had more reason to want vengeance than the thugs who constantly tormented Asgill at the tavern where he was incarcerated. "Worked upon by the pathetic appeals of Gordon," the Huddy family "became themselves supplicants in Asgill's favour," but their specific acts have not been determined.[104] Loved dearly by all who knew him, James Gordon must have had a natural charm.

After the vicious attack which nearly brought about Asgill's end, Gordon managed to secure a Newfoundland dog to protect him. And now that money was available to ease his confinement, there was money to feed the dog too. The dog became a faithful companion, and "slept constantly at the foot of his bed."[105]

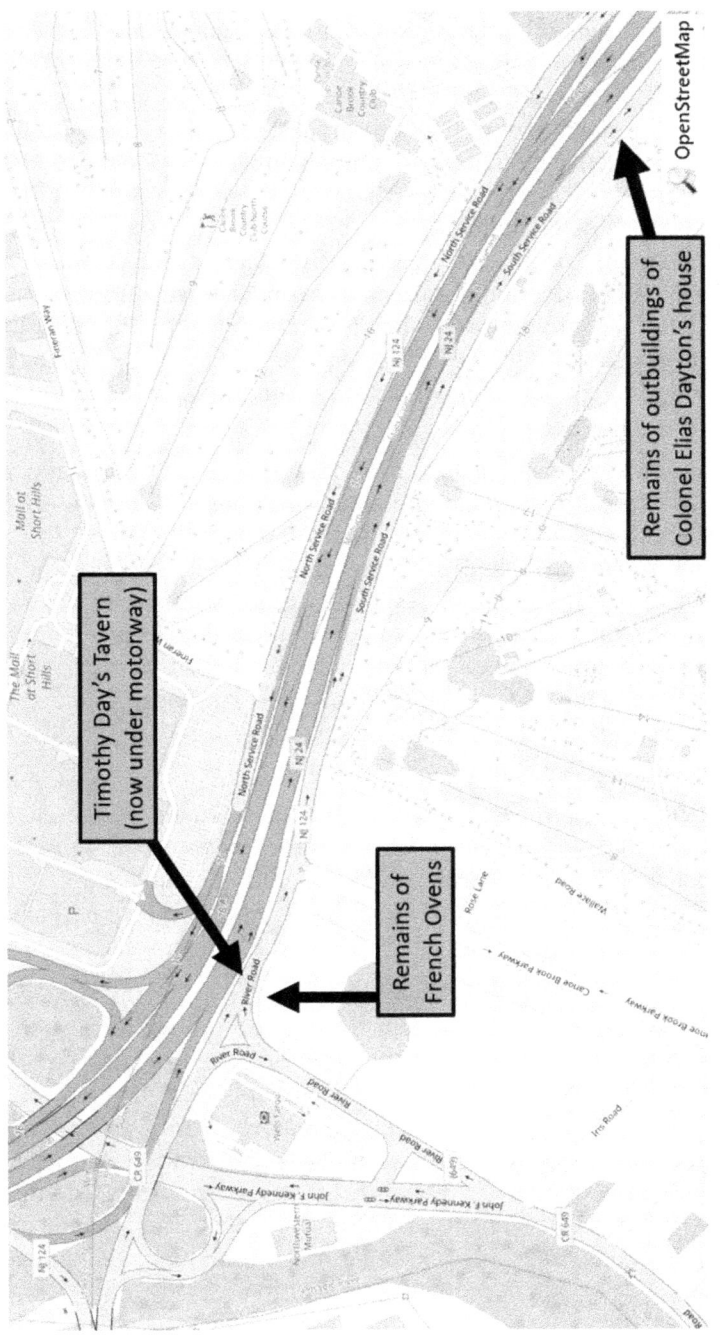

A modern map of Chatham, New Jersey, showing the 1782 locations of Timothy Day's Tavern and Colonel Elias Dayton's house on the Morris Essex Turnpike. Map from OpenStreetMap.org.

The Charles Asgill Affair

The humdrum of isolation was relieved occasionally by letters, including at least one in July from his fellow officer with whom he had drawn lots back in May, Captain Henry Greville.[106] In mid-August he received a visit from another of those comrades, Captain George Ludlow, who had "long been acquainted with his Family and Connections in England."[107] Ludlow was granted a pass by Washington to see Asgill "on motives of friendship and humanity"; he brought more letters.[108] None of these improvements meant that Asgill's life was saved. Nothing had changed the political situation during his time in Chatham. And for much of the time letters from home were being withheld from him, preventing him from knowing anything first-hand about the welfare of his family. But time was being bought, and if anything was going to save his life, it was time.

It wasn't until early in July 1782 that news reached London, not only via word of mouth, the press and the coffee houses,[109] but by official correspondence between New York and Whitehall.[110] James Jay, an American physician who had recently recanted his support for the rebellion and moved to London, responded quickly to the news. Jay's brother John was a former president of the Continental Congress and had just the month before arrived in Paris as an ambassador for peace negotiations; James Jay had also been involved in American politics, and penned an impassioned letter to Washington:

> I am now called upon to address you by one of those very affecting instances of private distress which deeply wound the heart; and excite all our tenderness in favour of the unhappy Sufferers. This melancholy Case is occasioned by the confinement of Capt. Asgill as an Object of retaliation; the news of which arrived here about six days ago.
>
> T'is a truly melancholy situation to be in, when the misfortunes of others affect our feeling, and prompt us to wish for their relief, at the same time that our reason tells us there may be insuperable obstacles to the exercise of this humane disposition. Unable as I am, in the present instance, to form a competent opinion on that point, I readily yield to the solicitations of friends, and the suggestions of my own feelings for the distressed family, to lay their condition before you; confident that your humanity and judgment will make a proper use of the information; and that your candour will put a proper interpretation on my conduct. My acquaintance with the family, and frequent visits to it in this hour

of affliction, enable me to give you the melancholy detail from my own knowlege.

Sr Charles Asgill, the father of the unfortunate youth, is very much indisposed, and in so critical a situation, having been of late severely threatned with an Apoplectic fit, that the News of his Son's confinement &c. is kept from him. His Lady, oppressed with grief, dare not see him, lest her behaviour might betray the secret, which would probably put an end to his existence. His Daughter, a lovely young Lady of great delicacy of mind and person, has ever since been delirous, raving almost perpetually about her brother, his execution and death. In this complicated Scene of woe, on account of her husband, Son, & daughter, Lady Asgill herself is sinking into a condition that is easier to be imagined than described.

Melancholy as these circumstances are, worse are still to be apprehended. Should the rigid hand of retaliation cut the thread of life of the unfortunate youth, the stroke would reach beyond himself. When I view the present distressing Scene, and think on what may happen, I feel, from the very bottom of my soul, for this amiable family. There is reason to fear that inadvertence or accident may carry the news of the Son's situation to the father, which would certainly precipitate him into great danger, if not immediately prove fatal: that agitation of mind may distroy the daughter: that complicated distress may sink the wretched mother. Yet nature may overcome the violence of the present shock: Hope may protract their wretchedness and their lives till they hear the Event: but should that prove contrary to their prayers and their hopes, the consequences will probably exhibit a Scene which humanity shudders to view even in imagination.

This, Sir, is the Case that looks up to you for relief. T'is the Case too of a friend; for Sir Charles has always been the decided and steady friend of our Country.[111]

Yes, as a staunch Whig, Sir Charles Snr (now on his deathbed), was most certainly a friend of the revolutionaries who were now threatening the life of his son. James Jay's letter was dated 19 July, but when Washington received it is not known. There is no record of any response.

The Charles Asgill Affair

By the end of the month the story of the drawing of lots and Asgill's captivity was in every newspaper. Some reports painted an optimistic picture:

> Although the situation of Mr. Asgill is not a pleasant one, there is every Reason to suppose it is not dangerous, as to going to his Life; for although General Washington seems determined to put an End to such Practices, by holding British Officers to answer for the Proceedings of those who act under English Banners, he is well known to possess too humane a Heart to suffer any Mischief to Capt. Asgill.[112]

But the public were deceived, had they believed there was any hope of Asgill's release.

The trial of Captain Lippencott concluded on 24 June, but the British legal process did not. Carleton, the commander in chief, had to review and approve the proceedings. Then, consistent with General Clinton's original promise of handling the affair, the Americans would have an opportunity to review the trial themselves. Carleton informed Washington that a copy would be sent, but not immediately: "the Minutes of the Trial, which are long, I shall order to be copied & sent to your Excellency."[113] It was not until 25 July that the proceedings were finally sent.[114] No mention was made of turning over Lippincott, who after all had been acquitted. As congressman James Madison wrote, expressing his dismay at how things were developing, "It is inferred that this murderer will not be given up, and consequently a vicarious atonement must be made by the guiltless Asgill."[115]

Asgill knew that this would be his death knell. He also knew that the time had now come to write to his parents, telling them of his deep sorrow to bring such dreaded news, and that he could only pray his father would forgive him for disobeying him, when, in his wisdom, he had forbade him to join the army and fight the American colonists. This letter was reported in the press, telling the world that he had written to apologise to his dying father that he had disobeyed him and gone to war, asking for his forgiveness. But would he, after his death, be prepared to grant his son his forgiveness?[116]

More time passed. By August, Washington had received and reviewed the proceedings of the Lippincott trial. The acquittal all but ensured that Lippincott would not be handed over for retribution. Washington was now, at least from his own original perspective, justified in ordering Asgill's immediate execution. But he had also had much time to reflect, to consider other factors. The case was now international news. Washington was the leader of the army, but not of the nation – he had a higher authority to answer to. He made a decision that had enormous impact on Asgill's fate. He sidestepped personal responsibility for the situation he had created by passing the entire matter on to Congress.

General Sir Guy Carleton, 1st Baron Dorchester, (3 September 1724 – 10 November 1808) Commander-in-Chief, North America. Unknown artist. Wikimedia Commons.

It was not until 26 August that case was laid before that governing body.[117] With the Asgill Affair in every newspaper, members of Congress surely had their opinions. But they were also busy with innumerable other matters, not the least of which was the possibility of finally negotiating peace after more than seven years of war. They had a nation to run, and a war to manage. Asgill, famous though he was, would have to wait his turn to be considered and debated in the halls of Congress.

The day before Congress received the case, General Washington wrote to Colonel Dayton ordering him to take his brigade to another location. He directed Dayton to "leave Capt. Asgill on Parole at Morris Town untill further orders."[118] Asgill apparently was taken to Morristown to new quarters. But on 5 September he was back in Chatham, where he wrote to Dayton that "the Inn at Morris is full & there is no lodgings for us, yet a while." He promised to return "as soon as we hear of any."[119] A week later he wrote again from Chatham; there is no evidence that he ever returned to Morristown.[120] In the first letter, he explained to Dayton that he interpreted Washington's directions that he be allowed parole in Morristown to mean "that every idea of retaliating upon me for the murder of Catp. Huddy was given up by his Excellency," and that he would be allowed to return home to "those, who must have long mourned my unhappy confinement." But during his brief stay in Morristown he learned that Washington had merely deferred his situation to Congress. He wrote to Dayton,

> How great & afflicting is the change; those pleasing Ideas are entirely vanished, & the prospect of continuing much longer in this dreadfull suspense will I fear if at a future time the decision proves favorable, be too late, to render comfort either to me or my aged Father.[121]

He closed with a plea that he had surely made before: "Permit to entreat you to intercede with Gen'l Washington in my behalf & to assist in relieving my present anxiety." His 12 September letter to Dayton was brief, pleading once more to be granted parole and allowed to return to England.[122]

There are stories of rescues being plotted, written in the nineteenth century, long after Asgill's ordeal. None are substantiated by information from anyone directly involved. A former British soldier who later met Asgill wrote:

> Captain Asgill had frequent opportunities of making his escape into New York; his whole guard (so greatly was he beloved by them) offered to come in along with him, if he would provide for them in England. Although these offers must have been very tempting to a prisoner, under sentence of death! yet he scorned to comply, as it would have involved more British officers in trouble.

Under Close Arrest

> He nobly said: 'As the lot has fallen on me, I will abide by the consequences.' ... Indeed the captain received very bad usage throughout his confinement; he was constantly fed upon bread and water. This harsh treatment constrained him to send his faithful servant to New York to receive and carry letters for him. This man ran great hazard in passing over the North River to New York.[123]

Another historian, citing no sources, wrote:

> One effort had been already made by the desperate partisan, single-handed, to rescue the young prisoner, while riding out on parole; and this was only defeated by Asgill's firm refusal to dishonour the pledge he had given his enemies. It was designed therefore to carry him away by force, which might easily have been done, so much license being allowed him in riding out for exercise, had not the communications of Parker (for such was the bold agent's name), put the keepers on their guard.[124]

After months of hardships, and never knowing when the call to the gallows would come, Asgill wrote again to Washington in the hope his plight might make him realise that his death would be a terrible stain on his reputation, and that of the emerging new American nation. On 27 September 1782 Asgill wrote:

> I hope when your Ex^y considers that I am not in the situation of a Culprit, that while on Parole I never acted contrary to the Tenor of it that my Chief motives for being so eager for further Enlargment is on account of my Family, these facts, I hope, will operate with your Excellency, to reflect on my unhappy Case, & to relieve me from a state, which those only can form any Judgment of, who have experienced the Horrors Attending it.[125]

Asgill was desperately worried about his father, since unreliable news reports and various rumours came through from guards and visitors. Some may have been based on newspapers that were notorious for publishing exaggerations and falsehoods, others from the vindictive visitors bent on tormenting him. At first he was told that his father was too ill to be told the news of his impeding execution, which was indeed the case. Then he was told by his guards that his father had died; they went so far as to address him as Sir Charles, implying that he had already inherited his father's title. He had no way of knowing how his family was coping, since their letters had been impounded by Washington, even though they were only designed to keep his spirits up, and nothing political had been addressed. His father had not died, but was extremely ill, remaining powerless, and ignorant of the situation, so could do nothing to help. An American newspaper repeated an item from a British paper:

> Sir Charles Asgill, having been judged passed recovery when the last express arrived from America, he has not been informed of the above melancholy situation of his son.[126]

Knowing how long it took for news to travel across the ocean, this must have caused the younger Asgill considerable worry.

Asgill also wrote to Washington, once more, on 18 October, ending his letter by saying "if your Excelly could form an Idea of my sufferings I am convinced the trouble I give would be excused"; so Washington most certainly had heard from Asgill's own hand that he had suffered during his confinement.[127] There is no evidence that, both times this had been brought to Washington's attention, he had investigated these allegations nor made any effort to establish if Asgill *was* being treated with the "tenderness" he had ordered.

In the meantime, inaccurate news was being reported in England as well as in America. In early August, London newspapers carried an optimistic item that was soon reprinted in other towns:

> We have the satisfaction of informing the public, from undoubted authority, that a letter is in town from a Member of Congress, bringing the agreeable intelligence that Capt. Asgill, who it was feared was doomed to suffer death, by way of retaliation, was (by the interference of Congress) discharged from his confinement.[128]

CHAPTER EIGHT

Intervention from France

Many of the officers who were surrendered at Yorktown in 1781 soon obtained paroles allowing them to return to Great Britain. One of them was Captain Charles Gould of the Coldstream Guards (sometimes referred to as the 2nd Regiment of Foot Guards), a longstanding friend of Captain Asgill. He left America soon after Asgill's "Unfortunate" lot had been drawn. Upon his arrival in England, Gould went to Richmond-upon-Thames to visit the Asgill family in what was then called Richmond Place, but is now known as the Palladian villa, Asgill House, also designed and built by Sir Robert Taylor, who had created so many of the iconic buildings, and carriage, commissioned for Sir Charles Asgill.

On arriving at the house he was shown into a room where Lady Asgill and her daughter, Amelia, were seated, and when he made the sad communication both ladies swooned away and fell, as it were, lifeless on the floor. Captain Gould immediately summoned a servant, whose surprise and horror may well be imagined when, on entering the apartment, he found the two ladies apparently lifeless on the floor, thinking that Gould had murdered them. Assistance and restoratives were quickly at hand, but the shock was necessarily great.[129]

Richmond Place, now known as Asgill House, from the front garden, 4 June 2002. Author's collection.

The Charles Asgill Affair

Drawing Room at Asgill House, where the family was informed of their son's impending execution. Author's collection, 4 June 2002.

Later on, Asgill would hear all about this from his family, but languishing in Chatham he had no idea that his mother and sister had been put through this ordeal. Amelia was inconsolable, and suffered a nervous breakdown, blaming herself for her brother's plight. A composer friend of the family (possibly Spanish, but his name is not known) took pity on Amelia and wrote a piece of music for her, entitled *Miss Asgill's Minuet*, no doubt intended to lift her spirits.[130] His mother, on the other hand, found fortitude; she knew that only she would have the strength needed to battle for his release.

While the news was slow to trickle through at first, soon it was being talked about all over Europe. Rumours swirled, became exaggerated, passed from the quayside to the coffee-houses, and from there to the press. Newspapers all over the continent carried the story – or versions of it – of the innocent young officer and gentleman condemned almost at random to suffer for a crime in which he had not had any part or knowledge.

Amelia Angelina (Asgill) Colvile (3 October 1757 – 12 July 1825), the eldest sister of Charles Asgill, 2nd Baronet. With kind permission of Jonathan Colvile.

Lady Asgill contacted the Home Secretary, Sir Thomas Townshend, who then, on 10 July, wrote a letter to General Carleton letting him know of his astonishment at Washington's plans to execute an innocent, conditional, officer. He informed the king, who in turn confessed his extreme regret at the news, but expressed his confidence in Carleton's ability to do what was needed.

> the time of extreme affliction is something which is represented to me as existing at present in that unhappy family makes it impossible for me to refuse in taking a step that may afford them any hope or consolation. M. de Rochambeau, and indeed Mr Washington too must expect that the execution of an officer under the circumstances of Captain Asgill must destroy all future confidence in Treatys & Capitulations of every kind, & introduce a kind of War that has ever been held in abhorrence among Civilized Nations.[131]

But this was not enough so far as Lady Asgill was concerned, so she wrote to the French Foreign Minister in Paris, Charles Gravier, comte de Vergennes, on 18 July 1782.[132] Though their country was at war with Britain, and France had been largely responsible for Britain losing the war and thirteen American colonies, still she knew he was the only one who could help her now. Her letter, written in her native tongue of French, has gone down in history as being the turning point in her son's salvation.[133] Her pleas for her son's life to be spared were heard in Paris, especially when Vergennes showed her letter to King Louis XVI and Queen Marie Antoinette. They were horrified that their ally, America, had violated the 14th Article of the Yorktown capitulation, and in executing the young officer, this deed would be done in the name of France as well as America.

The King, on behalf of the French nation, would have no part in it. He ordered Vergennes to write to Washington to tell him so. Vergennes did as ordered, on 29 July. Polished French diplomat that he was, he couched his words in the most civil tones, hiding the iron fist in the velvet glove, thereby providing Washington with an open door to do the right thing, while at the same time saving face. The punch-line of his letter went thus:

> There is one consideration, Sir, which, tho it is not decisive, may have an influence on your resolutions. Capt Asgill is doubtless your Prisoner but he is among those whom the Arms of the King contributed to put into your hands at York Town. Altho this circumstance does not operate as a safe Guard, it however justifies the interest I permit myself to take in this affair. If it is in your power, Sir, to consider & to have regard to it you will do what is very agreeable to their Majesties.[134]

Charles Gravier, comte de Vergennes (29 December 1719 – 13 February 1787) by Antoine-François Callet. Wikimedia Commons.

Though long, it was a brilliant letter. Right to the point, saying: do as you plan and you will be executing this officer in the name of France too. How could Washington possibly go forward with his promise of retribution?

The letter was signed and sealed right around the time that Washington received the proceedings of the Lippincott trial. But the letter had a long journey to make: from Paris to port, to the next ship bound for America, which then awaited favourable winds, followed by a voyage that

The Charles Asgill Affair

took six to eight weeks in ideal circumstances; at an American port, probably Boston or Philadelphia, the letter would be bundled with others bound for the same destination, and put in the care of a rider tasked with finding a commander in chief who frequently moved from post to post. All the while Asgill's fate hung in the balance, moving ever closer to the forefront of Congress's busy calendar.

Also in early August, London received news that Asgill was back on parole, and a free man, something which Asgill later wrote about, expressing his distress that this would mean his allies in London would think his danger of death had passed. The Home Secretary, Sir Thomas Townshend, appeared to think that the situation had been resolved; he wrote to Carleton on 14 August 1782:

> I had the honor of receiving on the 11th of last Month and laying before The King, your Letter No 4 of the 14th June, — with the several Inclosures relative to the unauthorized Execution of Captain Huddy of the New Jersey Militia by Capt Lippencott, and of the unprecedented and extraordinary resolution of Genl Washington with regard to Capt Asgill of the Brigade of Guards; and however unpleasing the subject may have been to His Majesty's Ear, He cannot too highly approve of the very judicious Measure you have taken thereupon, and rests in full confidence that the footing on which it appears to have been placed by your Letter No 9 of the 17th of June, that Justice has taken its course, and that the perplexing affair has come to a final decision.[135]

The time it took for correspondence to travel the Atlantic Ocean certainly played its part in muddying the waters. On the ground, in Chatham, it was certainly not over for Asgill.

It was late October when the Continental Congress finally took up the case of Captain Asgill. One might assume that by now the decision would be easy. But it was not. The vast majority of Congressmen were in favour of his execution, with only two lonely voices in dissention, namely the President of the Continental Congress, Elias Boudinot, and Congressman James Duane, a New York delegate. These gentlemen were vociferously against the murder of an innocent man. Boudinot wrote,

> A very large Majority of Congress were determined on his Execution, and a Motion was made for a Resolution positively ordering the immediate Execution — Mr. Duane & myself considering the Reasons assigned by the Commander in Chief

conclusive, made all the Opposition in our Power — We urged every Argument that the Peculiarity of the Case suggested, and spent three Days in warm Debate, during which more ill Blood appeared in the House, than I had seen — Near the close of the third Day, when every Argument was exhausted, without any appearance of Success, the Matter was brought to a Close, by the Question being ordered to be taken — I again rose and told the House, that in so important a Case, where the Life of an innocent Person was concerned, we had (tho' in a small Minority) exerted ourselves to the utmost of our Power — We had acquitted our Consciences and washed our Hands clean from the Blood of that young Man — That we saw his Fate was sealed —That we had nothing to do but request that the Proceedings should appear without Doors, as being equal to the Occasion, and the World should know that we had conducted the Measure with a serious Solemnity — That great warmth had been occasioned — Some harsh Language had taken Place — The Minds of Gent'n had been irritated — I therefore moved that the Question should be put off till the next Morning, on the Minority giving their Words, that they would not say another Word on the Subject, but the Question should be taken in the first Place, after the Meeting as of course — This was unanimously agreed to[136]

Three days of debate without unanimous agreement, a decision to put the matter to a vote, and a final plea by Boudinot to put off the vote until the next morning without further debate, official or private. The outcome of the vote was already clear; the ritual of querying each member of Congress for their yay or nay, and writing the proceedings into the official record, was but a formality. But it was a formality that allowed for one more twist, the sort associated only with novels or cinema, but which was very much real. Boudinot explained:

The next Morning as soon as the Minutes were read, the President announced a Letter from the Commander in Chief — On its being read, he stated the rec't of a Letter from the King and Queen of France inclosing one from Mrs. Asgill the Mother of Capt. Asgill to the Queen, that on the Whole was enough to move the Heart of a Savage— The Substance was asking the Life of young Asgill — This operated like an electrical Shock — Each Member looking on his Neighbor, in Surprise, as if saying here is unfair Play — It was suspected to be some Scheme of the Minority — The

President was interrogated, The Cover of the Letters was called for — The General's Signature was examined — In Short, it looked so much like something supernatural that even the Minority, who were so much pleased with it, could scarcely think it real— After being fully convinced of the integrity of the Transaction a Motion was made that the Life of Capt. Asgill should be given as a Compliment to the King of France — This was unanimously carried on which it was moved that the Commander in Chief should remand Capt. Asgill to his Quarters at Lancaster — To this I objected — That as we considered Capt. Asgill's Life as forfeited, & we had given him to the King of France, he was now a free Man, and therefore I moved that he should be immediately returned into New York, without Exchange — This also was unanimously adopted, and thus we got clear of shedding innocent Blood, by a wonderful Interposition of Providence.[137]

No matter how determined any member of Congress was to see Asgill hang, they demurred to the request of the French government, a government that had been the instrument of their successful war. Charles Asgill would not only live, but become a free man.

Even this required administrative procedures. It was not until 7 November that Congress passed a resolution to release Asgill on parole, as a gift from America to the French monarch. Then it was up to the army to effect his release. Washington was informed of the act of Congress, and in turn wrote of it to Asgill and to his detainers, on 13 November 1782.[138]

Confined in Chatham, Asgill knew nothing of this. He spent October in the limbo to which he had become accustomed at the tavern in Chatham, ignorant of his mother's exertions, of the intervention of the French government, the proceedings of Congress, of anything other than his impending death. But as night fell on 16 November there was one who sensed that something had changed. His Newfoundland dog, instead of sleeping at the foot of his bed as usual, lay across Asgill's body. The animal was far too heavy to consider sleeping with him on top, but repeated attempts to settle him elsewhere were futile. When the guards had moved on that night, for the first time they had left the door unlocked. The dog noticed what Asgill did not; thinking the unlocked door put his master in danger, he assumed a protective position, taking over the duty of the guards in securing his master's safety. No matter how often Asgill pushed him off, just as many times the faithful dog returned to his guard duties.[139]

Intervention from France

On Sunday morning, 17 November 1782, the young officer was handed the letter from Washington. "It affords me singular pleasure," it read "to have it in my power to transmit you the inclosed Copy of an Act of Congress of the 7th instant, by which you are released from the disagreeable circumstances in which you have so long been." Asgill surely felt far more pleasure than Washington, although the end of the affair was a relief to both of them for entirely different reasons. Washington astutely continued, "supposing you would wish to go into New York as soon as possible, I also inclose a passport for that purpose." That would have been enough, but Washington went on first to apologise for not responding to Asgill's letter of 18 October – "my not answering it sooner, did not proceed from inattention to you, or a want of feeling for your situation"; because he "daily expected a determination of your case, & I thought it better to await that, than to feed you with hopes that might in the end prove fruitless." With Washington's letter were delivered a number of other letters from his family, which had been withheld by Washington.

Washington's letter ended with what amounted to an apology, one that seems feeble given the injustice under which Asgill had been detained for six long months:

> I cannot take leave of you Sir without assuring you, that in whatever light my agency in this unpleasing affair may be viewed, I was never influenced thro' the whole of it by sanguinary motives; but by what I conceived a sense of my duty, which loudly called upon me to take measures however disagreeable, to prevent a repetition of those enermities which have been the subject of discussion—And that this important end is likely to be answered without the effusion of the Blood of an innocent person is not a greater relief to you than it is to Sir Yr most obt and hble servt.[140]

Relief flooded over him, and he was overwhelmed and almost weakened by this long-awaited news. Yet he felt torn in both directions because there was no doubt that the hardest part of his departure would be bidding farewell to the dear and loyal companion of his imprisonment, that astonishing man, Major James Gordon. Gordon was, after all, still a prisoner of war, allowed to leave his local parole in Lancaster only because of Asgill. How to thank him properly or enough? He did his best but, once more, as had been the case in Lancaster, his eyes must have filled with tears. Asgill implored Gordon to be sure to visit once he returned home, and he assured him he would certainly do so. There was one last favour needed, would Gordon please ensure that his canine companion was on the next available passage home? To which he happily agreed.

Asgill gathered his few possessions together and rode as fast as his horse would allow, heading for the British lines in New York, where he was to report to Sir Guy Carleton, who would be thankful to see him.

CHAPTER NINE

The Journey Home

As early as 4 June, General Washington had taken issue with the choice of Captain Asgill to suffer retribution for the murder of Captain Huddy – not because he objected to retribution, but because Asgill was a "conditional" prisoner, protected by the Yorktown capitulation treaty.[141] He wrote to General Hazen requesting that a different officer, a lieutenant from a Loyalist regiment, be sent in Asgill's place, "to remedy therefore as soon as possible this Mistake." But from Hazen's perspective no mistake had been made. Washington had, on 18 May, explicitly ordered Hazen to choose a conditional prisoner because no unconditional officer was known, by either Washington or Hazen, to be available; that information did not come to either of them until after Asgill's Unfortunate lot had been drawn.[142] And the other man in question was a lieutenant, not a captain. Washington's 4 June letter to Hazen had the tone of blaming Hazen for a mistake that was really a result of confusion about which prisoners were held in which places and on what terms. With prisoners held by different state governments in locations spread hundreds of miles apart, and no centralized system for managing them, it was the type of mistake that was bound to occur.[143]

Six months later Asgill was free, but for Washington the whole Asgill Affair was not about Asgill but about the murdered Captain Huddy. He had not yet proven that "my resolutions are not to be trifled with." With Lippincott acquitted there was no hope of the British command turning him over to the Americans. But there was another innocent person available to take Asgill's place.

Thirty-four-year-old Captain John Schaak of the 57th Regiment of Foot was captured in a skirmish at Sandy Hook on 26 May 1782 – one day before Captain Asgill drew his "Unfortunate" lot.[144] Captured in battle, Schaak was an "unconditional" prisoner; there was no treaty specifying terms for his confinement or release. He was held in "the Jersey Huts" in Chatham, in close confinement, and given no explanation of why he wasn't granted parole, at least locally, as was customary for captured officers. Was Washington holding Schaak as a surety in case Asgill was released or escaped?

The Charles Asgill Affair

When Asgill left Chatham he carried letters for General Carleton, among them a plea from Captain Schaak to intervene on his behalf.[145] That plea prompted Carleton to reason with Washington:

> The liberation of Captain Asgill was I trust founded on the equal principles of Justice and humanity, and I could wish Sir, that Captain Schaack was also released, not only to close a Question, of such intricacy that justice cannot act upon either side without losing its quality, but as there are also circumstances of ill health and infirmities, I presume not unknown to your Excellency, attending that Gentleman, which may render his confinement to both sides perhaps, equally unpleasant & unbecoming.[146]

Washington granted the parole, and Schaak was back in New York by early January 1783.[147] Apparently due to the "ill health and infirmities" mentioned by Carleton, Schaak resigned from the army in April.[148]

Washington wrote to Vergennes, the French Foreign Minister, a few days after Asgill's release. He seems to have been in a less than happy frame of mind that he had been deprived of revenge for Huddy's death. Many biographies of Washington describe how deeply pleased and relieved he was over Asgill's release, but this letter carries a tone of petulance, asserting that Asgill had no right to expect justice, with no mention that he was, in fact, a conditional prisoner protected under a treaty. Washington wrote:

> After I had the honor of receiving your Excellency's letter of the 29th of July, I lost not a moment in transmitting it to Congress, who had then under deliberation, the proceedings of the British Court Martial upon Capt. Lippencot, for the Murther of Capt. Huddy, and the other documents relating to that inhuman transaction — What would otherwise have been the determination of that Honorable Body I will not undertake to say, but I think I may venture to assure your Excellency that your generous interposition had no small degree of Weight in procuring that decision in favor of Capt. Asgill, which he had no right to expect from the very unsatisfactory measures which had been taken by the British Commander in Cheif to attone for a crime of the blackest dye — not to be justified by the practices of War and unknown at this day amongst civilized Nations. I however flatter myself that our enemies have been brought to view that

transaction in its true light and that we shall not experience a repetition of the like enormity.

> Capt. Asgill has been released and is at perfect liberty to return to the Arms of an affectionate parent whose pathetic address to your Excellency could not fail of interesting every feeling heart in her behalf.
>
> I have no right to assume any particular merit from the lenient manner in which this disagreeable affair has terminated — But I beg you to believe, Sir, that I most sincerely rejoice, not only because your humane intentions are gratified, but because the event accords with the wishes of his Most Christian Majesty and his Royal and Amiable Consort who by their benevolence and Munificence have endeared themselves to every true American.[149]

Washington was certainly right in denying that he was owed thanks for "any particular merit." Yet four years later he was incensed that none had been received. He had simply shirked his responsibilities and handed them over to the Continental Congress to decide.

Asgill made his way from Chatham first to Dobbs Ferry on the Hudson River and then by boat down the river to the city of New York. It was not a direct route, but was a typical way for people with passes to travel, avoiding dangerous areas on the front lines. Traveling alone, he was jeered and heckled, "expos^d to the insults of the people who wish^d to have seen me sacrific^d," since all chance of them taking vengeance for the death of Huddy, by his execution, now appeared lost.[150] Among the documents he carried was a report for General Carleton from Major Gordon, and Asgill must have appreciated the kind words his friend and advocate expressed concerning his conduct during confinement:

> Captain Asgill will have the honour to deliver this to Your Excellency, who is at last set at liberty by a Vote of Congress after a long and disagreeable confinement, which he bore with that manly fortitude that will for ever reflect honour upon himself. During the period that he was close confined he had frequent opportunities of making his escape, and was often urged to do it by anonymous correspondents, one of which assured him that if he did not make use of the present moment an order would arrive next day from General Washington that would put it out of his power for ever. This letter he gave me to read, and at the same time told me (that unless I wou'd advise him to do it) he never

The Charles Asgill Affair

wou'd take a step that might be the means of counteracting measures adopted by Your Excellency to procure his release, or might bring one of the officers of Lord Cornwallis's army into the same predicament, and that he had made his mind up for the worst consequences that cou'd happen from rebel tyranny.[151]

Also in Asgill's hands, besides his passport through American territory to New York, was a copy of Congress's resolution for his release. Although not its intention, the wording of the resolution makes clear the protracted timeline of the administrative process that culminated in his freedom:

> By the United States in Congress assembled November 7, 1782
>
> > On the report of a committee to whom was referred a letter of the 19 of August from the Commander in Chief, a report of a Committee thereon and motions of Mr Williams and Mr Rutledge relative thereto, and also another letter of 25 October from the Commander in Chief, with a Copy of a letter from the Count de Vergennes dated 29 July last, interceding for Captain Asgil.
> >
> > Resolved, That the commander in Chief be and he is hereby directed to set Capt Asgil at Liberty.
> >
> > Signed Cha Thomson Sect

Once Asgill arrived in New York he headed straight to the office of the Commander in Chief, in Manhattan, although he was not dressed, or even washed, for the occasion. He looked like the returning bedraggled prisoner that he was. Sir Guy greeted him warmly at his headquarters at Number One Broadway, Lower Manhattan.[152] Carleton expressed his great relief that his ordeal was now behind him. He explained that his main role had been one of stalling, by dragging out the matter of handing over Lippincott, to keep Washington's demands at bay.[153]

But Sir Guy spoke quickly and urgently, telling him that a packet boat bound for England, the *Swallow*, had just sailed from The Battery at the southern tip of Manhattan, a four-minute walk away from his headquarters. Packets were small, fast-sailing vessels that made monthly transits between America and Britain carrying mail and a few passengers. If Asgill was quick he might catch the *Swallow* while it was still negotiating the waters of New York harbour, before reaching the open sea. He was not at all prepared for a sea voyage, but this was an opportunity too tempting to miss. It was Wednesday, 20 November, and he had been free only since Sunday, but he could be on his way home if he could just catch that packet. For the first time in six months, things were happening fast for the young officer.

No.1 The Broadway, Old Kennedy House, where the British Commander-in-Chief's Headquarters was located in 1782 (cropped). Public domain.

The Charles Asgill Affair

Captain Asgill's movements within America, during 1781 and 1782.
These destinations are either verified, or there are indications that he may well have been there. It must be remembered that Asgill only became a *cause célèbre* following the drawing of lots in May 1782. Prior to that event, there are only very sketchy records of his movements as a young officer, within the Brigade of Guards, during his short time in America.

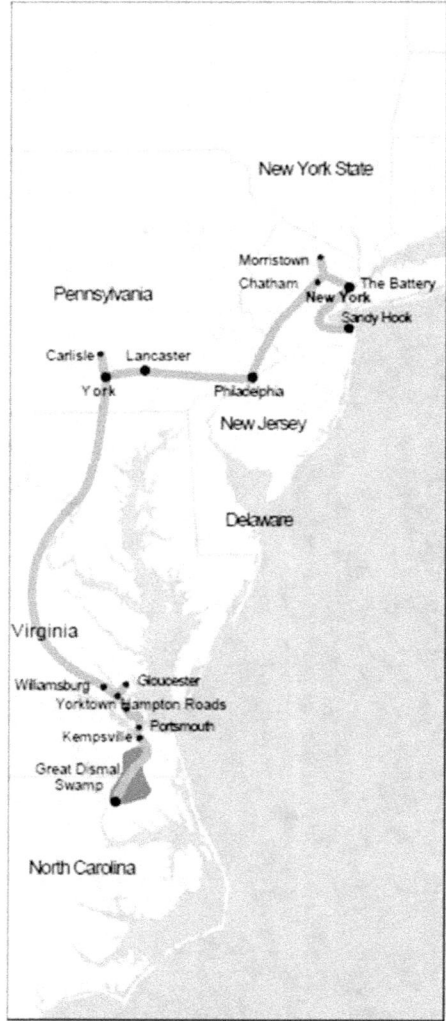

Hampton Roads, VA, where Asgill arrived, on board a transport ship, on or before 17 June 1781.
Portsmouth, VA, where he may have been under the command of Major General Alexander Leslie, for a short time.
Kempsville, VA, the military base for Leslie's troops, supporting Cornwallis's army, (which, at that time, nobody was sure of its location).
Great Dismal Swamp, VA/NC, (Black Swamp) where Asgill may well have been in battle, on 26 June 1781, giving him the opportunity to go to the rescue of patriot militiaman Colonel John Gregory.
Williamsburg, VA, where he first joined Lord Cornwallis's army for the four months remaining before Cornwallis's battles were over.
Gloucester, VA, from where Asgill saved boats, and men from drowning, on 16 October 1781.
Yorktown, VA, where he was with Cornwallis's army when he capitulated to the American and French armies on 19 October 1781.
York, PA, where Asgill was a prisoner-of-war and Major James Gordon was in charge of British prisoners-of-war at Lancaster.
Carlisle, PA, where he may (or may not) have attended William McCurdy's court martial.
Lancaster, PA, where Asgill drew the "unfortunate" lot of death on 27 May 1782.
Philadelphia, PA, where he stayed for a couple of days, on his way to Chatham.
Chatham, NJ, where he was under close arrest at Timothy Day's Tavern, for six months, awaiting death daily.
Morristown, NJ, where some reports state he was ordered to go, while awaiting the gallows, but Asgill refutes this, saying there was no accommodation available for him there.
The Battery, NYC, where he met Sir Guy Carleton at his HQ, but missed the *Swallow* which had already sailed from the Battery.
Sandy Hook, NJ, from where it is said that Asgill caught up with the *Swallow*, finally departing America on 20 November 1782. From whence he returned home on parole, arriving at his parents' house on 18 December.

The route covered by Asgill during his time in America 1781-2. By Alan Birch 2023. Copyright held by the author.

The Journey Home

British newspapers later reported his adventure succinctly:

> the Swallow Packet having sailed without him, he followed her in a Boat, but did not overtake her till she had got upwards of four Leagues to Sea. The Consequence was, that he came over without Servant or Baggage, and had but two Shirts on board.[154]

Later writers have elaborated on this passage without sources, saying that he procured a rowboat and rowed himself to the *Swallow*, others mentioning that he had the help of eight husky British tars to row after the sloop, even stating that he caught up with her from Sandy Hook and that Major Gordon accompanied him on a mad dash from headquarters to the docks, and so forth. The reality was probably less dramatic albeit feverishly fast; hiring a small, fast boat that could be sailed or rowed to a vessel only a few miles offshore would be a straightforward matter in a busy port like New York, and a wise commander of a packet would know to keep an eye out for signals to wait for last-minute emergency dispatches or passengers. Regardless of the details, Asgill got safely aboard and was on his way to England.

There were other officer passengers on board the *Swallow* including Captain William Carlyon Hughes of the 7th Regiment of Foot. Along with the vessel's commander, Captain Green, they no doubt saw to Asgill's comfort given his lack of baggage or preparedness for the voyage. Packet boats were swift, and prevailing winds always made the west-to-east Atlantic crossing quicker than east-to-west. After an uneventful passage of only 27 days, the *Swallow* put in to Falmouth, Cornwall on the southwest tip of England on 16 December. "We are very happy to communicate," proclaimed London newspapers that week, "the safe arrival of capt. Asgill on board the above packet boat – a circumstance which must give great satisfaction to the public in general."[155]

The route from Falmouth to London featured good roads and fast modes of transportation, so the mail brought by the *Swallow* made the 267 mile journey in only two days. With it came news of Asgill's arrival. A writer in the 1920s claimed that Asgill, too, arrived in London on 18 December. If this was the case, he arrived at the same time as news of his release. He knew that his parents would have received his farewell letter to them, written after he had believed there was no hope of salvation. The whole family would be in mourning. Katherine Mayo, without citing sources, claims that the *Swallow*'s Captain Green agreed to go to Richmond-upon-Thames with Asgill, arriving at his home before dusk on

The Charles Asgill Affair

18 December 1782. The ship's captain knocked on the door of the family home, Richmond Place. The servant, who answered the door, told him that Lady Asgill was receiving no visitors and shut the door in his face. He tried again, begging them to inform Her Ladyship that he bore good tidings; indeed her son was waiting outside, to be reunited with them. Once she realised she was not dreaming an impossible dream, she flew down the stairs at lightning speed, and her joy was complete.[156]

Regardless of whether that story is accurate, newspapers reported that Asgill appeared publicly in London on 25 December, Christmas day, for the first time since his release. He was "in pretty good health, considering what he was suffered in his confinement." They reported that "His legs are still swelled with the chains with which he was loaded. The rascal who guarded him in his prison beat him, because he complained of his servant having been ill treated."[157] But he was home. The nightmare was over. The Asgills were, once more, a family, just in time to enjoy Christmas together.

At church, they must surely have offered up their thanks to God.

Lady Asgill's immediate response is not known; eighteen months later she penned a letter that serves to show her feelings, in her own words. She wrote in French to Charles-Joseph Mayer, a Parisian author preparing a work called *Asgill, or the Disorder of Civil Wars* (*Asgill, ou les désordres des guerres civiles*), the first of several French publications regarding his experiences in America. The following year the letter was published in a newspaper, with the preamble,

> We have been favoured with the following translation of a letter from Lady Asgill to Mons. Mayer, author of the Romance, published in Paris some time since, under the title of *Asgill*, in which Captain Asgill's sufferings, as well as those of his family, are very affectingly described. We consider this a fresh proof of that elegance, spirit, and sentiment, which this amiable lady has displayed so often in the course of her delicate and heart-rending circumstances; and we doubt not but it will prove as acceptable to our readers as it has been pleasing to us.[158]

The translated letter read:

The Journey Home

London, June 1, 1784.

Sir,

 Illness has hitherto prevented my enjoying the singular pleasure of answering the letter you favoured me with the 26th of last April, which, as well as the ingenious work you have had complaisance to send me, clearly proves the full extent of the interest which the critical lot of my beloved son, during his imprisonment in America, has inspired you with. Deign to receive, Sir, the expressions of my sincere acknowledgment for a work that does so much honour to the goodness of your heart, and which shall be religiously preserved in my family. You have succeeded but too well in awakening my sensibility for those misfortunes my son has experienced, and which I participated without the means of rendering him any other comfort than my prayers for his life and liberty, prayers which were heard to effect, by the powerful and generous interposition of your august Sovereign, to whom I owe not only his, but my daughter's and my own existence. Yes, Sir, during the course of my life, as well as that of every individual of my family, we shall recollect with transports of gratitude, that we are indebted for life to the characteristic humanity of your Monarch. He is of the blood of the Bourbons, and that was an infallible presage for me. In fact, Sir, to diffuse happiness appears to be the whole ambition of your good King and incomparable Queen. They have truly made me the happiest of mothers in procuring the resurrection of my son, who was menaced with the sword of direful fate. My family shall offer up incessant vows for their preservation. May every day of their reign be as happy as that on which they changed our most gloomy despair into almost insupportable ecstasies of joy. Nor are we less earnest in our wishes for the comte de Vergennes, the rarest, as being the best of Ministers, who possesses all the virtues in so eminent a degree, and who was so ready in relieving my son by his tender and kind intercession; I must confess to you, Sir, that I could not read without the utmost agitation the following passage:

 "There (in France), a sage Minister holds with a steady hand the reins of Government; his genius is the powerful spring that puts every thing in motion. Europe beholds him. Europe respects him. Poor lady take courage! Your letter is gone. A man of feeling will read it – rest, unhappy lady – go attend your delirious daughter – calm the mind of your expiring husband – But alas! you cannot repose in the midst of so afflicting a scene!"[159]

I repeat, Sir, how much you have roused my sensibility, and renewed our sorrow. Be assured, Sir, that if ever you come to this island, my family and I will have the greatest pleasure in seeing you, in cultivating your acquaintance, and in paying you that attention which your feeling and sentiments claim of us. Then would we fondly dwell upon the panegyric of your beneficent Sovereigns – then you would hear our acknowledgments to that rare Minister, and witness the lively sense we entertain of that merit which you have so well described. Accept, Sir, the assurances of my family's esteem, as well as the sentiments of gratitude, with which I have the honour to be, &c. Theresa Asgill[160]

CHAPTER TEN

Paris; to thank Their Majesties

Asgill was not the only one to be thankful that the traumas of his Chatham experience were now behind him. Peace talks to end the war, which at this time involved France and Spain as well as the newly-independent United States, were underway in Paris. Had he been executed, the impact on these talks may have been considerable. When hearing of Asgill's release, John Adams, leading the American negotiations, wrote to Robert R. Livingston: "The release of Captain Asgyll was so exquisite a Relief to my feelings, that I have not much cared what Interposition it was owing to — It would have been an horrid damp to the joys of Peace, if we had heard a disagreeable account of him."[161] Certainly, had Asgill lost his life, the prisoner-of-war, the comte de Grasse, would probably have lost his too. He had been captured by the British at the Battle of the Saints on 12 April. News of the battle, which resulted in de Grasse's capture and the loss of seven French ships of the line, reached London on 18 May.[162]

The best and fastest way for people to learn of Asgill's return home, and maybe get a look at England's new celebrity, was to attend the King's Levée on 25 December 1782.[163] This Levée was what would be termed an "Open House" in modern terminology, but essentially meant a daily moment of intimacy and accessibility to King George III. The great and the good could turn up at the royal palaces and mingle with everyone else doing the same thing. It was delightful for him to see old friends again, many of whom served with him in the American War. All those he encountered had seen reports of his experiences in America, reported in *The London Chronicle* twice weekly (amongst others) which had broken the news on 16 July. All had feared for his life and all expressed great joy at the final outcome. There is no doubt that Asgill was a popular officer and a dear friend to those who knew him, as his mother had said in her letter to Vergennes "My son (an only Son) and dear as he is brave, amiable as deserving to be so".[164]

He was also hearing strange stories about his own experiences. While he was still in captivity, newspapers were publishing an indistinguishable mix of truth and rumours. As early as 5 September London newspapers were reporting Asgill's release and attributing it to "the Interference of the Count de Rochambeau," commander of the French forces that had cooperated with Washington at Yorktown.[165] By 16 December the press offered accounts of Asgill having been taken to the gallows three times, and a gibbet erected outside his cell window in preparation for his execution, with a notice nailed to it stating "Up Goes Asgill for Huddy."[166] He was surprised that the rumours had begun so soon, well before his return, but he was determined that he would ignore these reports since, as he later wrote:[167]

> the extreme regret with which I find myself oblg^d to call the attention of the publick to a subject which so peculiary if not exclusively concerns my own Character & private feelings will induce me to confine what I have to say within as narrow a Compass as possible

There was nothing to be gained by entering this unsavoury rumour-mill; or worse, to be labelled as the instigator of the rumours. Besides, his feelings were raw, and he needed to maintain a distance from those experiences, now, which was paramount. His life and potential death had been discussed in public for quite long enough and he did not want to fuel the fire.

One expectation of him would not be fulfilled. He knew George Washington would be awaiting a bounteous missive of appreciation from him, thanking him for all that he had done to secure his release. But he knew he had been saved by French intervention, and he did not feel that his conscience would allow him to *genuinely* thank a man who had so illegally threatened to take his life, and who had apparently done little himself to relieve him (apart from constantly bewailing the need to have condemned him to death in the first place). Washington had even failed to investigate the bad treatment Asgill received, when Asgill had written to tell him of the "Horrors" and "sufferings" attending his confinement, on 27 September and again on 18 October. He had written to him from Chatham to explain that if this situation continued, it would soon be too late to save him from death in captivity, so close had he come to meeting his Maker while confined at the tavern. His silence on all fronts was the best he could offer. Four years later, he wrote:

> I leave for the public to decide how far the treatment I have related deserv^d acknowledgements – the motives of my silence were shortly these: The state of my mind at the time of my release was such that my judgement told me I could not with sincerity return thanks; my feelings would not allow me to give vent to reproaches[168]

Paris, to thank Their Majesties

It was France where thanks were due and now would not be too soon to commence preparations and requests needed to fulfil his great desire to thank Their Majesties for having taken pity on his poor mother and her son. His regiment, the 1st Foot Guards, was posted in London and he was once again an active duty officer, but he requested a leave of absence to visit France.[169]

He was given the permission needed and would happily have gone to Paris as soon as he knew his presence would be convenient to Their Majesties. However, this would not be possible at the moment, because the visit would have to include Lady Asgill; how could he possibly go without her? She had fallen ill before his return home. The stress she had been under, carrying the burden alone, had now caught up with her and she was confined to bed by her physicians. Rest, they assured her, would restore her health.

It did, but it took a year, so they could not travel to France until October 1783. Four of them set off, Her Ladyship, Amelia, Maria and Asgill. They had to leave behind his father, the 1st Baronet, since he was still far from well, and little Caroline, only just nine years old the month before their departure, was too young for the sort of company they would be keeping. Caroline stayed at home to help look after their father.

By 20 October they had arrived in Paris, and British newspapers duly reported their progress, telling readers that "They are to set out immediately to Fontainbleau, to return their Thanks to his Excellency Comte de Vergennes."[170] They met Gilbert du Motier, Marquis de Lafayette and Armand Louis de Gontaut, duc de Lauzun, both of whom had been opponents of Asgill in America. Lafayette commanded American forces in Virginia at the time of Asgill's arrival there in June 1781, and then a wing of the army that encircled the British at Yorktown; Lauzun commanded a legion of cavalry and infantry in the French army that collaborated with the Americans at Yorktown. Now these former enemies were hosts. At the Palace de Fontainebleau, the royal residence southeast of Paris, Lafayette and Lauzun presented Asgill, his mother and his sisters to King Louis XVI and Queen Marie Antoinette. It was also here that Charles Asgill finally met in person the comte de Vergennes, the Foreign Minister who had effected his reprieve.[171] "The gratitude of the one, and the generosity of the other, are pleasing instances in favour of modern manners," wrote another correspondent.[172]

The Charles Asgill Affair

Queen Marie Antoinette and King Louis XVI on 7 February 1775 at the Château de la Muette. Detail from painting by Josef Hauzinger. Wikimedia Commons.

A correspondent who saw the family, whose letter subsequently appeared in the newspapers, wrote of Asgill, "This young man is of a very good figure, but short," and after discussing his ordeal in America asserted, "He must be possessed of great courage and force of mind, not to have sunk under a situation so oppressive and violent, and an imprisonment which continued three months" (correction, six months). He described Asgill's sisters as "fine women," singling out Amelia:

Paris, to thank Their Majesties

"especially the eldest, who is beautiful as an angel. She has not as yet recovered from the nervous attack, caused by her apprehensions for her brother's life, upon hearing his cruel destiny, which effected her the more, as it was in consequence of her particular solicitations that his father consented to his entering into the guards, and going to America."[173]

Having the opportunity to be presented to King Louis XVI and Queen Marie Antoinette, and to thank them personally, was an occasion that stayed in all their memories for the rest of their lives. The monarchs were kind, gracious and hospitable to them throughout the several months the Asgill family were in their country. Who could possibly have imagined they had been enemies such a short time before? Sarah, Lady Asgill, being of French heritage, helped enormously, and, of course, she had ensured her children were all able to communicate in French. Their private audience with Their Majesties at the Palace de Fontainebleau was truly magical and the family thanked them both profusely for intervening with Washington, which resulted in Charles Asgill's life being saved.

None of the Asgills left a record of their visit, at least not one that has been found. A friend of theirs had this to say about her own first introduction to Queen Marie-Antoinette. "I was so frightened that I could hardly walk or know where I was. I made three curtesies. I then tried to speak to the Queen, which is the etiquette, but I found I had no voice. She spoke to me directly and was so gracious and good-humoured that she reassured me". On meeting her again, at Fontainebleau, she was watching the high spirited Marie-Antoinette at a gambling card game called lansquenet; and on another joined the wild boar hunt in the forest where the young Queen, clothed in a blue and gold habit, rode a small black horse with an immense velvet saddle. She noted that the Queen "looked beautiful and rolls herself about upon her horse and is quite at ease. She had no rouge, and looked so well I could not take my eyes off her."[174]

Asgill's sisters were almost certainly equally overcome when meeting the beautiful Queen. But Their Majesties' hospitality went beyond that, and they welcomed the Asgills at Versailles and Fontainebleau on several occasions to celebrate the young man's return to his loving family. In 1783, George Montagu, 4th Duke of Manchester, was appointed Ambassador to France to "supervise the conclusion of treaty negotiations between Great Britain and France, Spain, and the Netherlands;" on 3 September he was a signatory to the Peace of Paris at Versailles for Great Britain to end the American War of Independence.[175] Vergennes was another of the signatories. Manchester would undoubtedly have been at

The Charles Asgill Affair

the centre of the Asgills visit to Paris. His vivacious wife, Elizabeth Dashwood, would also have looked after the family well.

A rather unexpected surprise was to be told of a forthcoming play in Asgill's name, *Asgill, Drame, en Cinq Actes, en Prose*; written by Jean-Louis Le Barbier the younger and dedicated to Lady Asgill. Sarah Asgill was particularly tickled that the play was dedicated to her. Quite right too! Le Barbier was better known as a painter than a playwright. When the play was finally published in London and Paris in 1785, he wrote in the introduction, "Obstacles that I was able to foresee, and motives that I needed to respect, having prevented the launching of this Play in 1783, I had decided to put it to one side, until I was convinced by public interest that the subject should be presented in the French Theatre." He continued with a lengthy discussion of his anxieties, not only on the subject matter, but his feelings of inadequacy in writing it. A translation of the introduction appears in Appendix Six.

Sarah Asgill's handwritten copy of Le Barbier's play, dated Paris, 1783, has been kept to this day by the descendants of Asgill's sister, Amelia. It has always been their belief that his play was performed for them in Paris, by command of Queen Marie-Antoinette, but there is no evidence that it was and Le Barbier's own introduction indicates that it was not. Nevertheless, Le Barbier had done a good job and the Asgills were pleased with his efforts. In 1784 Lady Asgill carried on correspondence with Le Barbier, also presented in Appendix Six.

When the time came to return home, the family would likely have been sad; they'd had a wonderful time enjoying the beautiful city to the full, but time had flown by. Their itinerary is not known, but their time was most likely spent being entertained, reciprocating kindnesses shown to them, visiting *l'Opéra* and *l'Théâtre*, soaking up the culture and beauty of the city, and marvelling at the magnificent *Cathédrale Notre-Dame de Paris*. The Asgills had been there a few months, but it was time for Charles Asgill to return to his military duties in London. His faithful Newfoundland dog, having long been reunited with the family in England after surviving a sea passage from America, would have been delighted to see them home.

CHAPTER ELEVEN
James Gordon; 3rd Laird of Ellon

In March 1782, when Major James Gordon was still in Lancaster, Pennsylvania, as a prisoner of war, he discussed the burial of a fellow officer prisoner from the 80th Regiment who had died. One of the regiment's officers asked about burial in the churchyard. "No," said Major Gordon; "you see that spot near the barracks where so many British soldiers have been buried; that is the place where I myself should wish to lie were I to die, and there you will deposit the remains of this British officer; for you know that officers and soldiers should not be separated, and at the last day the soldiers would be greatly surprised if they saw no officer."[176] Did he have a premonition that he would die in America?

After he had departed Chatham, with Asgill in November, Gordon returned to captivity in Lancaster where he remained until June 1783. At that time, peace negotiations having been concluded, the prisoners were released and returned to the city of New York, which was still the British headquarters in North America. He was promoted from major to lieutenant-colonel, and had an active role in administrative activities associated with the repatriated prisoners of war. He presided over a lengthy court martial that heard numerous cases concerning residents of New Jersey, where he had been so dearly loved by both Americans and British alike that all wished to have their cases heard by him, knowing him to be so fair in his judgements. But he was, in the words of a fellow officer, "quite an altered person" after his experience with Asgill, "having lost much of the liveliness of disposition which has always seemed so natural to him."[177]

Becoming increasingly ill, during his time in New York, dropsy (an old-fashioned or less technical term for oedema) finally brought him to his deathbed, where he was visited by Captain Samuel Graham, who had served with him since their arrival in America in 1779. Graham was one of the thirteen officers to draw lots in Lancaster back in May 1782, and was also a member of the court martial. Gordon handed Graham a letter he had received from Lady Asgill, in which she expressed her deepest gratitude for all that he had done for her son. Gordon asked him to thank her for the letter, and to apologise to her for being unable to respond. This Graham did, once he himself returned to Britain. Lady Asgill's letter to Gordon was written soon after her son's return from captivity, when she

The Charles Asgill Affair

was so ill herself, but, of course, she did not know that the recipient was also too ill to respond. She wrote:

> Sir — If distress like mine had left any expression but for grief, I should long since have addressed myself to you, for whom my sense of gratitude makes all acknowledgment poor indeed; nor is this the first attempt; but you were too near the dear object of my anguish to enter into the heart-piercing subject. I earnestly prayed to heaven that he might not add to his sufferings the knowledge of ours. He had too much to feel on his own account, and I could not have concealed the direful effect of his misfortune on his family, to whom he is as dear as he is worthy to be so. Unfit as I am at this time, by joy, almost as unsupportable as the agony before, yet, sir, accept this weak effort, from a heart deeply affected by your humanity and exalted conduct, as heaven knows it has been torn by affliction. Believe, sir, it will only cease to throb in the last moment of life with the most grateful and affectionate sentiment to you. But a fortnight since I was sinking under a wretchedness I could no longer struggle with. Hope, resignation, had almost forsaken me. I began to experience the greatest of all misfortunes, that of being no longer able to bear them. Judge, sir, the transition, the day after the blessed change takes place — my son is released, recovered, returned, arrived at my gate, in my arms! I see him unsubdued in spirit, in health, unreproached by himself, approved of by his country, in the bosom of his family, and without anxiety but for the happiness of his friend, without regret but for having left him behind. Your humane feelings that have dictated your conduct to him, injured and innocent as he was, surely must participate in every relief and joy his safety must occasion. Be that pleasure yours, sir, as every other reward that virtue like yours, and heaven can bestow. This prayer is offered up for you in the heat of transport as it has been in the bitterness of my anguish; my gratitude has been soothed by the energy it has been offered with; it has ascended the throne of mercy, and is, I trust, accepted. Unfit as I am, for nothing but sensibility so awakened as mine could enable me to write, exhausted by too long anxiety, confined at this time to a bed of sickness and languor, yet I could not suffer another mail to go without this weak effort. Let it convey to you, sir, the most heartfelt esteem and gratitude of my husband and children. You have the respect and esteem of all Europe, as an honour to your country and to human nature, and the most zealous friendship

of, my very dear and worthy Major Gordon, your ever affectionate and obliged servant, (Signed) T. Asgill.[178]

Mary Ellen Gordon, married John Balfour, 5th of Balbirnie, on 23 July 1771. Sister of Lieutenant Colonel James Gordon. Artist unknown. Courtesy of Robert Balfour, the Lord Lieutenant of Fife.

The Morris-Jumel Mansion in Upper Manhattan where James Gordon died in 1783. Wikimedia Commons.

A few years later Asgill wrote a letter about how he felt regarding this very special man:

> the Consolation & Support which I deriv^d from the exalted Friendship & Kind Compassion of Major Gordon of the 80th Reg^t who feeling for the distresses of a Brother Officer, by the best of Hearts in the cause of humanity & unwilling to leave a Youth of eighteen unadvis^d & unsupported to act in so peculiar & difficult a situation, sacrific^d every Comfort to partake my hardships & Confinement & by the impulse of his excellent & noble Heart, felt on the [first] acquaintance all the steady & persevering Zeal that the longest & most tried Friendship could hope for or Claim he was there the whole of this transaction the partaker of my hardships the support of my Spirits, & the monitor of my conduct I am delighted at having the opportunity of proclaiming to the World his generous & benevolent attentions Tho whilst I do

justice to his Memory I aggravate the sensations of regret, which I must ever retain, for the loss of him[179]

Gordon passed away on 17 October 1783 at the age of 48.[180] He died of dropsy at his quarters at the Morris House, near the British post at Kingsbridge in Upper Manhattan. On the 18th his coffin was carried from the Morris House to Trinity Church, Lower Manhattan; a distance of 13 miles. So loved was he by all who knew him, that this was thought the best way of showing him that respect. He was buried in an unmarked grave, with military honours. The loyalist Bishop Charles Inglis, the Rector of Trinity Church, or one of his subordinates, recorded his name as "L. C. Gorden" in the church burial register, rather than as it should have been, "Lieutenant Colonel James Gordon." Consequently, has anyone ever known this astonishing man is buried there?

In 1938 an Oxford Professor of History, Keith Feiling, wrote a review of *General Washington's Dilemma* by Katherine Mayo. He titled it "Only one hero – Major James Gordon."[181] Feiling wrote that Mayo's book is, in some ways, a novel and as such deserves a hero. He maintains that the hero is not Washington, not the young Asgill and not the murdered Huddy. The hero, he declares, was Major James Gordon. Pointing out the many ways in which Gordon helped to change the disastrous path chosen for Asgill, he highlights that he worked to death to save his young charge. He helped Asgill face up to the fate handed to him; shared all the hardships of his imprisonment; "spurred Frenchmen and British and Americans to action" and, in doing so, shamed them all to eventually do the right thing. He says that Gordon's "plain courage and humanity shines in this ugly, tangled business." He thinks that all readers of Mayo's work should be grateful to her for bringing yet one more hero to their attention – "as the curtain falls on her intensely-wrought, moving, brief, and rounded tragedy."

James Gordon's legacy is that of a Hero. It is no less than he deserves. This is Mayo's description of him:

> James Gordon, First Major of the Eightieth Foot, was, like his regiment, Scot to the core. Middle-aged, tall, rather heavily built, one glance at the man revealed his character. Frank as the sun and as friendly, brave to a fault, and as generous, utterly self-forgetful in the face of others' needs, neither the activity of his mind nor the dignity and cheerfulness of his spirit would bow to the worst of days. His soldiership, as displayed in the two past years of campaigning, had won him Lord Cornwallis' special praise –

coupled once with an aside: 'When I first knew Gordon, twenty years ago, gay in gay London, who could have guessed how much lay in the man?'[182]

This is believed to be the room in which James Gordon died, in the Morris-Jumel Mansion. Author's collection, 28 May 2019.

He is no longer the forgotten hero, buried in an unmarked grave at Trinity Church, New York City. A stanchion has been erected with the wording:[183]

James Gordon

1735-1783 (Buried in an unknown location within Trinity Churchyard)

Lt. Col. in the British Army during the Revolutionary War. In 1782, Gordon was a central figure in the Asgill Affair, a military dispute in which General George Washington ordered the

James Gordon; 3rd Laird of Ellon

retaliatory hanging of an innocent British prisoner-of-war, Capt. Charles Asgill.

Gordon advocated for Asgill's release and volunteered to take his fellow officer's place at the gallows. Neither man was hanged.

The Asgills' visit to Paris had been wonderful, except for the terrible news of Gordon's death. His passing was announced in British newspapers in late December 1783,[184] and had a dreadful impact on them all.

> **James Gordon**
>
> 1735-1783 (Buried in an unknown location within Trinity Churchyard)
>
> Lt. Col. in the British Army during the Revolutionary War
>
> Gordon was a central figure in the Asgill Affair, a military dispute in which General George Washington ordered the retaliatory hanging of an innocent British prisoner-of-war, Capt. Charles Asgill.
>
> Gordon advocated for Asgill's release and volunteered to take his fellow officer's place at the gallows. Neither man was hanged.

A memorial stanchion erected at Trinity Church, New York City, in honour of Lt. Col. James Gordon. Erected on 8 March 2022 (239 years after his death). With kind permission of Trinity Church, NYC.

In his Memoir, Samuel Graham had this to say about his dear friend, James Gordon. "The eloquent remarks by General Burgoyne, on the death of

General Frazer, may appropriately be transferred to the memory of my lamented friend Lieutenant Colonel Gordon".

> To the canvas, and to the faithful page of a more important historian, gallant friend, I consign thy memory. There may thy talents, thy manly virtues, their progress, and their period, find due distinction, and long may they survive long after the frail record of my pen shall be forgotten.[185]

CHAPTER TWELVE

The False Rumours

Once back in London in 1784, following their lengthy visit to Paris, Asgill's military duties were resumed. He went on recruiting service, tasked with enticing likely men to enter into careers as soldiers. The attrition of wartime was over, but even during times of peace a steady flow of recruits was needed to keep regiments up to strength. Because peacetime demand was less than during war, recruiting officers could be more selective than during wartime. Each regiment did its own recruiting; although recruiters were free to go to whatever places they thought good men might be found, the 1st Foot Guards recruited almost entirely in England.[186]

His men had always shown confidence in him and his experiences in America meant his name was well known. This may have improved his chances of attracting men to volunteer for service in the Guards, but he didn't much care for his newfound notoriety. He had become a *cause célèbre*, since all of Europe had heard his name and had followed, with interest, the manner in which developments had transpired. The Asgill Affair would be remembered for the rest of his life, and, in America at least, for centuries following his death, and while he could dine out on the tales told, he would have preferred anonymity.

On 12 November 1786, at the age of 29, his eldest sister, Amelia, married Sir Robert Colvile at his parents' residence in Portman Square. Robert Colvile's country estate, Hemingstone Hall, one of Suffolk's loveliest houses, is an exceptional example of a Dutch-gabled Jacobean manor house set within lovely gardens. It is Grade 1 listed and contains fine panelled rooms and some vivid Jacobean wall-painting. Its grounds, of historic importance, include a listed walled garden by Lanning Roper, sunken parterres, a woodland garden and gazebo.[187]. Asgill would have spent some happy times visiting them there, after they had both settled into a comfortable married life. Their wedding was such a happy occasion, but little did he know that four days later Washington would once more impact his life.

Totally unfounded rumours concerning his imprisonment in America, continued to circulate throughout Europe. The story of General Rochambeau interceding on Asgill's behalf, the tale of a gallows erected outside a jail cell window, allegations that Washington himself deliberately instigated the harsh treatment Asgill received, were still being told years after his return. A member of Congress, Elias Boudinot, wrote:

> Capt. Asgill soon sailed for England, and on his Arrival, he behaved without any sense of Obligation for his Escape by suffering the most false and injurious acc's of his Liberation to be published in all their Newspapers without an attempt to contradict them — Indeed I found generally, that the British Officers did not think themselves bound to keep their Word or perform Acts of common Gratitude & Generosity with Rebels.[188]

Asgill still refused to be drawn into any of this, never confirming nor denying anything. He felt that it was not his place to comment, since the grapevine had born fruit months before his return to England. Besides, his state of mind, whenever his thoughts wandered back to his time in Chatham, ensured that he would never voluntarily speak of those days, outside of his family and close friends. He had made the treatment he received clear to Washington himself in two letters written whilst still a prisoner in Chatham. In his 18 October letter he wrote: "if your Excell^y could form an Idea of my sufferings I am convinced the trouble I give would be excused."

But in 1786, four years after the events, an anonymous correspondent wrote a letter to one of Washington's former aides-de-camp in America saying that London coffee houses were abuzz with accusations against Washington, and that Asgill himself was the source of those accusations. No matter that the stories had begun to circulate in London when Asgill was still in captivity; the years had erased any memory of when the tales began. Word of the accusations soon reached Washington, and he believed that Asgill was assailing his honour.

Another of Washington's former aides, Colonel David Humphreys, passed on for publication a very carefully manicured bundle of Washington's correspondence relating to the Asgill Affair, including nothing of a damaging nature to his master. Not included were Asgill's letters of 27 September and 18 October, in which he told Washington of the "horrors" and "suffering" he was experiencing, and of his "fear fatal consequences may attend much longer delay." Washington had known that Asgill had been mistreated, but chose not to include that information in his published correspondence. Also omitted was Washington's letter of 18 May explicitly ordering protected officers to be considered for execution.

False Rumours

Humphreys made the arrangements for publication without mentioning that Washington was involved in selecting the specific documents to be published, and those to be omitted.[189] Of the four letters that Asgill wrote to Washington, only one was included in the collection that ran in the *New Haven Gazette and Connecticut Magazine* of 16 November 1786, under the banner headline: "The Conduct of GENERAL WASHINGTON, respecting the Confinement of Capt. Asgill, placed in its true Point of Light."[190] This ensured that the truth would be carefully shielded from public gaze.

The only letter from Asgill to Washington published in the *New Haven Gazette* was written on 17 June 1782, just three weeks into his confinement. He had been housed by Colonel Dayton for little more than a week, if that, and in this letter he had told Washington that he had been treated kindly, which was the case up until that time. Four years later, this one letter was presented as proof that that same kindness had lasted for the next five months, which had certainly not been the case. Having thanked Washington for ordering that he be treated with tenderness, his own words have been used, and twisted against him, ever since. The absence of Asgill's letters to Washington of 27 September and 18 October, and of Washington's letter of 18 May, has skewed the narrative for two and a half centuries. It was a cover-up, and Asgill was enraged that the story had been told only through Washington's eyes. Washington's letter of 18 May was also not included in *The Writings of George Washington* published in 1891, the major primary source on Washington used by twentieth century historians.[191]

When copies of the *New Haven Gazette* material reached England, Asgill was beside himself with anger. This carefully curated newspaper account of his experiences bore little resemblance to the truth. He sat down immediately to write a letter pointing out that the "anonymous American Correspondent" whose information had reached Washington about the rumours circulating in coffee houses and the European press had absolutely no basis in truth whatsoever. Besides, before deciding to believe the rumours, his "correspondent" could have spoken to Asgill personally to clarify the situation, since he was living in London at this time. On being charged as the perpetrator of the false rumours, Asgill wrote:

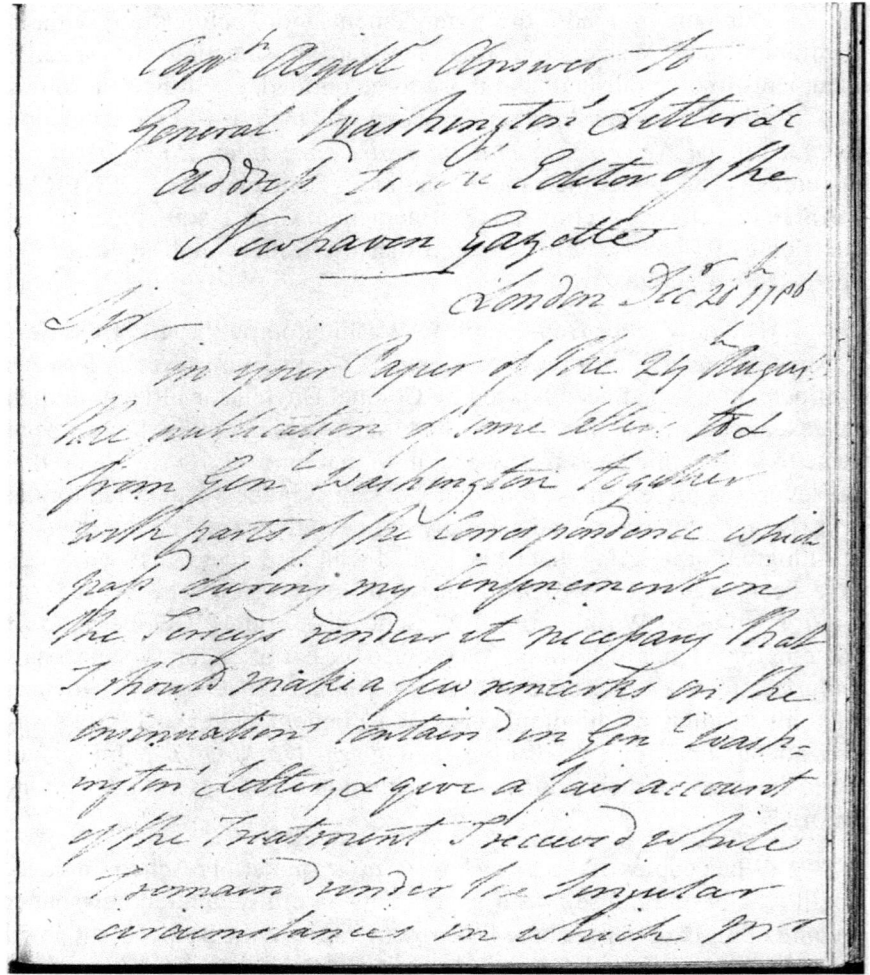

First page of Asgill's Letter of 20 December 1786 to the editor of the *New Haven Gazette*. Wikimedia commons.

I find myself obliged to call the attention of the publick to a subject which so peculiary if not exclusively concerns my own Character and private feelings will induce me to confine what I have to say within as narrow a compass as possible – Very little is necessary to the anonymous letter of the American Correspondent, who boasts his introduction to Coffee house sages & making his

False Rumours

assertions on coffee house authority so confidently affirms that charges were exhibited against General Washington by Young Asgill of illiberal treatment and cruelty towards himself ... It is sufficient to say that this Gentleman whoever he is never took the pains of ascertaining the truth of the intelligence he received from his Coffee house sages by an application to me, tho I almost resided constantly in London[192]

Asgill also took the opportunity to outline the appalling treatment accorded him at the tavern, about which he had never spoken in public. His letter was dated 20 December 1786 and was addressed to the editor of the *The New Haven Gazette*, requesting him to publish his response to Washington's accusations. The editor, Josiah Meigs, a man with close connections to Yale University in New Haven, was an avowed revolutionary himself, and was determined not to do anything to damage Washington's reputation. Asgill's letter would have done Washington a great disservice, had it been published when he wrote it, especially when he made it known that he had, indeed, been treated exceedingly badly. The rumours were not right: as Asgill pointed out, he never went to the gallows three times, and never had a gibbet erected outside his cell window – but the truth was no less damning than the rumours. Most probably the deciding factor, for Meigs, was that Asgill declared that his life had been saved by Their Majesties, the King and Queen of France, and definitely not by Washington.

After his papers had appeared in print, Washington wrote to David Humphreys to thank him for making the arrangements to publish them. In this letter of 26 December 1786, he wrote that "The manner of making it was as good as could be devised," terminology that suggests Washington colluded with his aides to exclude Asgill's letters of 27 September and 18 October, along with Washington's to Hazen of 18 May.[193] Asgill's reputation as a cad and a liar has stuck steadfastly to him ever since then, but George Washington's reputation has remained intact. Although the famous "cherry tree" story is a myth, the statement which goes with it, "I cannot tell a lie," has travelled, undamaged, throughout time. That was the intention of the author of the myth.

Extract from Washington's letter to David Humpreys, 26 December 1786. George Washington Papers, Library of Congress.

Josiah Meigs did not publish Asgill's letter, whether of his own volition or by consultation with David Humphreys and George Washington. It remained virtually unknown until 233 years after it had been written, when it was published in the Winter 2019 issue of *The Journal of Lancaster County's Historical Society*. The editor of that Journal wrote: "Little time elapsed between the publication of Col. Humphrey's article defending General Washington in the *New-Haven Gazette* and Captain Asgill's response, but the editor decided against publishing Asgill's response."[194]

For two centuries, most authors relied solely on the version of events presented in *The New Haven Gazette* or variations of it, propagating the smear on Asgill's reputation. In the early 1800s one American writer did boldly defend Asgill's reputation, using first-hand knowledge of his character to put him in a more favourable light. It was Alexander Garden, one of Asgill's two American friends from Westminster School, so long ago. Garden served in the American War, but on the opposing side from Asgill; he was an officer in a cavalry regiment and aide-de-camp to a senior American general. In spite of this, he could not help but refute the assault on the character of his old friend. In his account of the Asgill Affair, he addressed the rumours and put blame for them where it belonged:

> To what then, but the deadly animosity of a nation, instigated by the successful opposition to their arms, and the threatening prospect of the loss of empire, can be attributed the falsehoods and scurrilities with which the British prints, on both sides of the Atlantic overflowed. Their editors unblushingly

asserted "that Captain Asgill was thrice conducted to the foot of the gallows, in order to complete the threatened retaliation; and, moreover, that the instrument of punishment erected in front of his prison, did not cease to offer to his eyes, the dreadful preparations, more awful than death itself."

The promulgation of these calumnies, could not fail to make a deep impression where the truth was not known, and, with a poignancy unspeakable, to lacerate and afflict the bosoms of his friends and family.[195]

After fixing blame for the rumours on the press, and animosity over having lost the war and the American colonies, Garden went on to express his own reaction to them, pointing out that he knew Asgill personally:

Notwithstanding so satisfactory a termination of this eventful business, the British Gazettes continued lavishly to disseminate abuse, and even to assert, "that Captain Asgill himself, was, on all occasions, loud in proclaiming the unnecessary rigour extended towards him by General Washington, and a scandalous want of delicacy on the part of the American officers, with whom he came in contact." I was greatly surprised at these statements, and loth to believe them. I had been a school-fellow of Sir Charles Asgill, an inmate of the same boarding-house for several years, and a disposition more mild, gentle, and affectionate, I never met with. I considered him as possessed of that high sense of honour, which characterizes the youths of Westminster in a pre-eminent degree.[196]

Rather than make his own conclusions blindly, Garden discussed the matter with a mutual friend who had also been a classmate at Westminster:

Conversing sometime afterwards with Mr. Henry Middleton, of Suffolk, Great Britain, and inquiring, if it was possible that Sir Charles Asgill, could so far forget his obligations to a generous enemy, as to return his kindness with abuse. Mr. Middleton, who had been our contemporary at school, and who had kept up a degree of intimacy with Sir Charles, denied the justice of the accusation, and declared, that the person charged with an act so base, not only spoke with gratitude of the conduct of General Washington, but was lavish in his commendations of

The Charles Asgill Affair

> Colonel Dayton, and of all the officers of the Continental army, whose duty had occasionally introduced them to his acquaintance. It may now be too late to remove unfavourable impressions on the other side of the Atlantic, (should my essay ever reach that far,) but it is still a pleasure to me, to do justice to the memory of our beloved Washington, and to free from the imputation of duplicity, and ingratitude, a gentleman, of whose merits I had ever entertained an opinion truly exalted.[197]

It was a superb defence, but too little, too late. Garden's account was not published until 1828, by which time Asgill had been dead for five years, and numerous histories of the war were in print that carried the conventional opposing view. And in many ways Garden's account mirrored others, reprinting the letters from Lady Asgill, Vergennes and Washington written in 1782; it was easy to read it and not pay particular attention to the important statements about Asgill's character. Moreover, although Garden spoke well of Asgill and attempted to put things right about the rumours, he nonetheless wanted to sell books. In early America, disparaging George Washington was not the way to do that. Garden closed his discussion with a parting shot that seems intended to bring Asgill down a notch in the eyes of readers, thereby not placing him above Washington in terms of honour or esteem:

> Some further particulars relative to Captain Sir Charles Asgill, having come to my knowledge since making the above statement, I think myself called upon by imperious duty to publish them, however decided their tendency to destroy the favourable sentiments I wished to inculcate of his candour and veracity. The prepossessing traits of character that adorned his early years, I can never forget, nor is it possible for me to suppose, that to Mr. Middleton, whose entire family (with a single exception) were enthusiastically engaged in the service of America, he would have expressed a grateful sensibility for favour shown him, while in the circle of his more intimate associates, he had industriously propagated sentiments so decidedly contradictory. One circumstance, I confess, not only wounds my feelings, but staggers my faith.
>
> That no reply was made to that highly interesting and pathetic letter of General Washington, informing him of his liberation from captivity, and freedom from the penalties that threatened his life, accompanied at the same time by passports,

False Rumours

which enabled him to join his companions in New-York, and speedily to assuage the tumults of his mother's breast, and restore his sister to reason and to happiness, must appear strange, and in nowise consistently with propriety, to be accounted for. It manifested (to give it no harsher name) a want of politeness and respect, that with a gentleman, must be deemed impardonable. If my opinions have been more favourable to him than they ought to have been, I sincerely lament it, since in the language of the poet I can truly say, "I hate Ingratitude more than the sin of lying."[198]

Given that Asgill was initially held captive and condemned illegally, being a conditional prisoner of war and therefore protected from such actions, he can hardly be blamed for not sending a letter of thanks to Washington for releasing him. It is equally clear that Garden was not apprised of the sufferings Asgill *had* experienced when under close arrest in Chatham. Washington was, after all, the architect of Asgill's captivity and condemnation, and it was Congress that ordered Asgill's release at the behest of the French foreign minister. Garden, in an effort to elevate the exalted Washington, suggested that he should be thanked for doing no more than following the orders of Congress.

When she researched her own book on the Asgill Affair, Katherine Mayo was perplexed at finding no evidence of Asgill's own account of how he had been treated, fully expecting that he would have made the truth known to the world. She wrote:

> But Asgill himself, be it because the story escaped his notice, be it because the facts had been such that additions of embroidery could scarcely make them worse, or be it because he knew thus early the eternal folly of story-fighting, seems to have published no statement at all concerning his American experience.[199]

Asgill kept his counsel for four years, but his temper snapped when confronted with the half-truths and manipulated account presented to the world by his arch nemesis. Perhaps it is not surprising that Washington remains his nemesis to this day – his tactics ensured that his reputation remained intact, whereas the picture he painted of Asgill has stuck with him, like a bad smell, for two and a half centuries.

Washington himself, in unrelated correspondence, inadvertently summed up the general perception that has clung to Asgill since his imprisonment: "it is well known, that when one side only of a Story is heard, and often repeated, the human mind becomes impressed with it, insensibly."[200]

CHAPTER THIRTEEN

Troubles in America Now Behind Him

Asgill spent a decade in London, after returning from America, performing his duties with the regiment in whatever capacity he was called upon. Although modest about his experiences during the war, he was sometimes highly visible. In the procession for the King's Birthday in June 1785, for example, where the wealthy and famous often showed off their new carriages, "Capt. Asgill appeared in a neat chariot, with a stripe, in imitation of the Prince's carriages of last year."[201] His celebrity assured that he was occasionally the topic of gossip that the press pounced on, as in March 1785 when a London paper ran the notice, copied in other papers, "A few nights since a young lady, famed for her beauty and fortune, made a matrimonial trip towards Scotland, from Saint James-palace, with Capt. Asgill of the Guards."[202] It would be intriguing to know who that young lady had been.

In early 1788 he had the honour of being appointed as Equerry to His Royal Highness, Frederick Duke of York and Albany, King George III's second son. He held this appointment for 35 years, forming a close relationship with his royal master during that time. The newspapers reported: "Captain Asgill kissed hands, on being appointed Equerry to his Royal Highness the Duke of York."[203]

But that year was a sad one for the family since his father's health deteriorated badly. He passed away on 15 September at his home in St. James's Square.[204] They buried him on the 21st at St Bartholomew's-by-the-Exchange, under the Great Stone in the North Quire. It must have been a very well-attended occasion, with the then current Lord Mayor, John Burnell, arriving in the Golden Coach Asgill's father had commissioned in 1757. Several earlier holders of that office would have likely also attended. Perhaps even some members of staff at the Mansion House, who still remembered him, were also present, along with many friends from the banking community. His charities would also have supported the family with their attendance at this sad time, since they all meant so much to him.

One of the most important attendees was Sir Robert Taylor, the sculptor and architect who had designed the 1st Baronet's home, Richmond Place, which had been completed in 1767 as a weekend retreat for the family. Tragically, Sir Robert "got a cold by attending the funeral of his friend Sir Charles Asgill, which turned to a mortification in his bowels" and died six days later, on 27 September.²⁰⁵ The Asgills must have been deeply shocked to hear of this tragic end to such an outstanding career. Taylor is commemorated with a plaque dedicated to him on the wall of the south transept of Westminster Abbey.

Prince Frederick, Duke of York and Albany (16 August 1763 – 5 January 1827) by Sir Thomas Lawrence. Wikimedia Commons.

The 1st Baronet had been a stalwart head of the family; one who had been dearly loved by them all, and his loss naturally had a big impact on the forever doughty Lady Asgill (henceforth to be known as Dame Sarah Asgill). With her husband's death, now her only son inherited the Baronetcy and was thereafter styled Captain Sir Charles Asgill, 2nd Baronet – the new young head of the family, just 26 years old.

The 1st Baronet's obituary in the *Gentleman's Magazine* stated that "he was a strong instance of what may be effected even by moderate abilities, when united with strict integrity, industry and irreproachable character," going on to say "it is said he has died worth upwards of £160,000 the principal part of which devolves to his son, Capt. A, of the first regiment of guards."[206]

The 1st Baronet had written his will from Cork Street, Burlington Gardens in the County of Middlesex, on 11 April 1782, just after his son's twentieth birthday.[207] Little would he have dreamt, when he wrote it, just what a calamity was about to befall his family in the coming months. The second Charles Asgill was already a prisoner-of-war in America, which probably eased the father's concerns over his safety now that fighting had ceased, but the 1st Baronet's health was a huge concern to the family very soon afterwards. This might have prompted him to put his affairs in order.

A transcription of this will runs to fourteen pages of closely typed text. He set up a trust to ensure that his son would benefit from his lands and estates, but would be unable to sell them since they should be passed to his heirs or to his siblings. The old Sir Charles was shrewd, and while generous, wanted his estate to be handed down to his grandchildren, or if not, back to his daughters and their families.

> during the life of my said son Charles Asgill in trust to preserve the contingent uses and estates hereinafter limited from being defeated or destroyed and for that purpose to make entries and bring actions as occasion shall require, but nevertheless to permit and suffer my said son Charles Asgill and his assigns to have and take the rents, issues and profits thereof to and for his and their own use during his life and as, to, for and concerning such of the said premises as are copyhold in trust for my said son and his assigns for and during the term of his natural life; and from and immediately after the decease of my said son Charles Asgill then, as, to, for and concerning all and singular the said Premises hereintofore devised both freehold and copyhold with their

respective appurtenances to the use of and in trust for the first son of the body of my said son Charles Asgill, lawfully to be begotten ... and for default of such issue then to the use of and in trust for my three daughters Amelia Asgill, Maria Asgill and Caroline Asgill equally to be divided between them, share and share alike ... My son Charles, by my Will, is to have all my Real & Copyhold Estates ... In case of a qualification for a member of parliament the above freehold estates are more than sufficient, as the sum required is £300 per annum[208]

He always had aspirations for his son to become a Whig Member of Parliament, which he much preferred to an army career. With no heir involved in banking, the firm of Asgill, Nightingale & Nightingale became John, William & George Nightingale, and survived only until 1796.[209]

Two months after her husband's death, a man came to Dame Sarah's house to sweep the chimney. This sort of thing was not unusual at a fashionable London home, but moments after the sweep departed the alert butler noticed that a silver cross-stand was missing. He pursued the sweep and discovered the cross-stand hidden under his waistcoat. The thief was arrested and tried at the Old Bailey on 10 December, just fourteen days after the crime. The sweep, William Goodman, was found guilty and sentenced to "transportation," that is, sent to a British colony overseas, for seven years.[210] The cross stand was valued a three pounds.

Asgill's younger sister, Maria, had been ill for a long time and they all knew they would lose her in the foreseeable future. On Christmas Day, 1788, their first without their darling Papa, Maria went up to her room and wrote her will. She had no witnesses and no legal advice, but she wrote what was in her heart and made the bequests she wanted to make. She was now a wealthy woman in her own right, following her father's passing. After making her bequests she wasn't sure what would remain to donate to charity; she simply wrote that the money "left unemployed" must be donated, but did not name any particular charity.

After inheriting the Baronetcy our young Asgill became exceedingly wealthy, with properties in London and farm holdings in Waltham Abbey, Essex, from which he obtained a healthy income. He was not very good with money, and got into bad habits of gambling at the London clubs he frequented. It nearly led to disaster for him, much to the anger of his sisters, their father's will having stated that if he died childless then his inheritance would revert to them and their descendants. They were not at all happy about the way of life he had adopted. On occasion he would be with his master, the Duke of York, who was also over-fond of gambling. After hours he could be found enjoying London's high life, with its gaming tables; he was perpetually in debt because of his excessive gambling on cards and racehorses. Asgill's situation became quite serious, and his gambling habits were noted and commented on. William Knollis, 8th Earl of Banbury (later General Knollys) wrote to his mother, wife of the 7th Earl of Banbury:

> Sir Charles Asgill is clearly a great match for Miss Ogle but he is in the high road to ruin, owing many thousands, and his connection with the D of Y does not contribute to lessen his expenses – plays deeper than most young men though pretty successful – his fortune has been good, amounting to 50 thousand pounds when his father died – is near about 27 years of age.[211]

Knollis had no doubt heard about Asgill's brush with money-lenders, which ended up in court as "Asgill v. Chandless" in 1790. The case was brought by Asgill against Thomas Chandless, and centres on a dispute between the two over the terms of redemption of two loans (in the form of annuities) extended by Chandless to Asgill in May 1788, and also the scale of charges by Chandless against Asgill for arranging the loans. There is no evident record of a court decision on the dispute.[212] He had only come into his inheritance two years earlier. Hardly a surprise that his sisters were unhappy.

But he was so happy! He had met the most wonderful woman, who he knew would be his wife the moment he met her. Her name was Jemima Sophia Ogle, the youngest daughter of Admiral Sir Chaloner Ogle, whose seat was Worthy Park House in Martyr Worthy, Hampshire. She was frequently in London, staying with married siblings, so that he had ample opportunity to squire Sophia to balls, dinners and assemblies; they became an integral part of London high society, and at Royal functions as well.[213] At this time he also became more interested in politics and the two of them became part of Georgiana, the Duchess of Devonshire's Whig set. The duchess was a formidable and fascinating woman and they never failed to enjoy time spent in her company, whether it be at political fundraisers, balls, concerts or the theatre.

As was common among the aristocracy of her time, Georgiana routinely gambled for leisure and amusement. Her gaming spiralled into a ruinous addiction, however, made worse by her emotional instability.[214]

The Charles Asgill Affair

However cross Asgill's sisters were with him, it is easy to see how he was influenced by the company he kept.

On 3 March 1790 Asgill was promoted to command a company in the 1st Foot Guards, with the rank of lieutenant-colonel.[215] This was one of his most pleasurable appointments since there was nothing he enjoyed more than being with the men of his regiment.

Georgiana Cavendish, Duchess of Devonshire (7 June 1757 – 30 March 1806) by Joshua Reynolds. Wikimedia Commons.

His dear younger sister, Maria (Harriot Maria), died at Dame Sarah's Portman Square residence on 30 April 1790. She was 22 years old and should have had her whole life ahead of her. While the family grieved her loss, it quickly became clear that there would be problems ahead regarding her un-witnessed will. Whether it had been her intention or not, £2,000 was left "unemployed." This, in later years, caused great divisions within the family. The youngest sister, Caroline, was at the time of Maria's death not quite 16. Caroline married Richard Legge a decade later, on 5 April 1800. The matter of Maria's will was taken to court by them. The verdict went against them and, after appealing, the judgement read: "There never has been a case in which the executors have been permitted to take the residue for their own use."[216]

Asgill's beautiful, quirky, flirtatious, and very outspoken Sophia (charms which had so quickly endeared her to him) were engaged by 1789. As she became part of the family, she became as attached as the others to the Newfoundland dog that had protected Asgill so well during his imprisonment in America. But "to the great grief of his mistress and all the household," the dog was stolen.[217] How such an enormous hound could have been taken against its will is hard to comprehend, but the faithful companion was gone.

Asgill fared much better with Sophia. They were married at St Swithuns, Martyr Worthy, on 28 August 1790.[218] Sophia's Ogle family welcomed their son-in-law wholeheartedly, and the couple spent time with them at Worthy Park House. Asgill formed a close bond with her younger brother, Charles, who had joined the Royal Navy three years earlier, and who went on to have a highly successful career with that service. Fighting in the French Revolutionary and Napoleonic Wars, he was eventually promoted to Admiral of the Fleet in 1857. Sophia's father, Admiral Sir Chaloner Ogle, 1st Baronet, had fought in the revolutionary war in America, so the two men would have had plenty to talk about. Asgill loved them all, her brothers and sisters alike, and they all became friends for life – society column inches and diaries covering their get-togethers. Family meant so much to Sophia that her family became his family too. The following year, 1791, they moved into the perfect home for them, back in London, at 6 York Street, which became their home for the whole of their married life.[219] They became exceedingly attached to it, even though Asgill's army postings prevented them living there as much as they would have liked.[220]

On 5 November 1791, along with Lady Malmsbury, Sophia was appointed Lady of the Bedchamber to Princess Frederica Charlotte of Prussia, who on marriage became the Duchess of York and Albany.[221] As soon as her household was established Sophia commenced her duties, so then both Sophia and Charles Asgill had roles to play within the royal household. The Duchess of York was another dog-lover, so Sophia would have been in her element had she been asked to help with the royal dogs. The famed artist John Hoppner RA (4 April 1758 – 23 January 1810) painted the Duchess's portrait to celebrate her engagement to the Duke of York, and one of her Pomeranian dogs is shown in that painting. Sophia is sitting at her feet, wearing a fetching green velvet gown. Sophia was godmother to Hoppner's granddaughter, Helen Clarence, so she may have been instrumental in the choice of artist.[222]

Sophia, Lady Asgill, née Ogle (1770 – 30 May 1819). Asgill's wife, and Lady of the Bedchamber to the Duchess of York. Photographed and cropped from the original Hoppner portrait by the author.

Sophia delighted in arranging a superb wardrobe for herself, now that she had royal duties to perform, and the newspapers reported on one of her gowns:

> Lady Asgill was by far the most elegant and best dressed Lady at Court: she wore a white satin dress superbly ornamented with gold and white velvet. Over the petticoat hung a wreath of oak leaves and acorns, and the bottom was trimmed with a rich tassel fringe. Her Ladyship looked remarkably beautiful.[223]

The Duke's marriage to Frederica was not a happy one, unfortunately, and Frederick and Frederica separated. The Duchess then retired to Oatlands Park, Weybridge in Surrey, where she lived eccentrically until her death. The Duke even installed his mistress, Mary Anne Clarke, close by for a while, so there was never any question of reconciliation.[224] High-stakes gambling is reported to have taken place at Oatlands, where Frederica died in 1820.

The Asgills took on a house maid named Sarah Paris in January 1791. She worked for them until 11 June when she was discharged as the Asgills prepared to travel, but they graciously allowed her to stay at the house until she found a place of her own. When they returned to the house some time later, she had moved out. Gradually, Asgill and his servant began to notice that things were missing; in his words, "I missed things every day, as I wanted them. A book, *Bell's British Theatre*, volume 14. Three shirts. Two handkerchiefs. A leather case. A toothbrush case. Another book, *The Political Works of Lord Littelton* [who was a Tory turned Whig]. A glass decanter. A glass goblet, a glass tumbler, two wine glasses, a desert knife, "and a piece of linen cloth, containing two yards and a half." Suspecting Paris, Asgill obtained a search warrant and, accompanied by two colleagues, found and searched her lodgings. There they found the missing household items. The shirts, identifiable because they were marked with Asgill's initials, were in possession of a man who sometimes lived with Paris; he happened to be a servant to Colonel Banastre Tarleton, another famous officer from the American War.

Sarah Paris was brought to trial on 7 December 1791. A court official told the court, "this poor woman seems very big with child, and she begs to sit down," a request that was granted. Attending court, Asgill testified to the facts of her service, the missing items (which he identified when they were produced in court), and the process of finding them. He closed his testimony by saying, "I am induced to think it was extreme poverty which drove her to it, for she is with child, and has another, and the supposed father was absent; I had the best character with her." Of Tarleton's servant, he said, "My lord, I beg leave to observe, that I was induced to be lenient to this man, in hopes that he will take care of the children."

The court found her guilty. Before sentencing, Asgill made one final statement to the presiding magistrate: "My lord, I beg leave to recommend her strongly to mercy." The court was clearly moved by Paris's sad circumstances and influenced by Asgill's kindness. The judgement was:

> Sarah Paris, you have been indicted and convicted of a felony, in stealing a quantity of linen, and other things, the property of Sir Charles Asgill, your master, who has very humanely recommended you to mercy, as also the Jury have recommended you with equal humanity; your situation influences me to pass on you the mildest punishment that I can pass upon you; and as I have a power, by the late act of parliament, to commute burning in the hand for a pecuniary punishment, my sentence is, that you be fined 1 shilling and discharged.[225]

The French Revolution must have been a dreadful sadness to the Asgill family and, on 21 January 1793, one day after being convicted of conspiracy with foreign powers and sentenced to death by the French National Convention, King Louis XVI was executed by guillotine in the *Place de la Revolution* in Paris. His Queen, Marie Antoinette was also guillotined, on 16 October 1793. How could the family not have mourned the very people who had saved their son? They would surely have reminisced over their Paris vacation, when they met both monarchs.

Britain was a major combatant in the French Revolutionary Wars. In late 1793 Asgill was posted to the Continent and joined the army there under the command of His Royal Highness, the Duke of York, to whom he was Equerry. The Flanders Campaign (or Campaign in the Low Countries) took place from 20 April 1792 to 7 June 1795, and was the inspiration for the well-known children's song about the "The Grand Old Duke of York" who marched his men up and down hills. Asgill's friend from his days in America, one of those thirteen officers who drew lots, Samuel Graham, was in Flanders at the same time. He later wrote:

> On the 1st February 1793, France declared war against England. The result of this declaration was the embarkation of the

Duke of York, with 10,000 men, for Holland, where his force was augmented by a reinforcement of 25,000 Hanovarian, Hessian, and Brunswick troops. The first events of this campaign were most encouraging; and at one time 130,000 men might have been assembled to march upon Paris – a movement which would probably have brought the war to a satisfactory termination. History can only now point to the campaign as a monument of the fatal influence of cabinets in war. England, over-elated at the first successes, and more attentive to her maritime interests than to the general objects of the war, insisted on his Royal Highness laying siege to Dunkirk, the possession of which was much coveted. This unfortunate resolution sealed the fate of the campaign. A force, which in one mass was capable of achieving great things, was thus fractioned into part, exposed to ruin in detail. The national indignation of the French was roused at the bare idea of the soil of France being touched by foreign hosts; and in the national enthusiasm thus created, the rival parties which had risen out of the revolution, overlooked their party animosities, and united in the most energetic measure to resist the invasion of their country. The fearfully disorganized state of society at that time was also highly favourable to the morale of the French army, by introducing into its ranks great numbers of a superior class; the profession of arms, strange to say, being one which offered more security than any other. General Foy, in describing the mode of fighting of the republican armies, says – "Sometimes, in the midst of a shower of balls, an officer, or a soldier, or a representative of the people, would sing the hymn of victory; the general displayed his hat, with its tricoloured plume, on the point of his sword, as a rallying point; whilst the men advanced at the charge, their drums beating, and their voices rending the air with cries of 'Forward, forward; success to the republic!'"[226]

The following newspaper report gives a glimpse of Asgill's experience in Flanders:

British Head Quarters, Arnheim Nov 27, 1794. Never, since the commencement of this war, has there been such a dearth of intelligence, as at the present period. All is report, and the gossip has been very busy for these few days past. Report says, that the British Parliament is pro-rogued. This information was conveyed to the army by a flag of truce from the enemy! The enemy have a

The Charles Asgill Affair

camp near Neimegen; and we continue to have six regiments on the opposite bank of the Waal, to keep them in countenance. The firing in the environs of Tiel is at intervals very brisk. The French continue to bombard Port St.André, but with little effect. They are raising very strong Fortifications at Port Loevestein where the Merse and the Waal meet. The Dutch, our brave and loving allies, continue to give us the most spiriting proofs of their affection. One of the Burghers of this town slapped the door in Sir Charles Asgill's face, when he shewed his billet, on the arrival of the grenadier battalion in this town. Lord Cavan's servants were threatened, if they dared put his horses in the stables allotted to them by the Magistrates. In short, there is not a British officer, but has been insulted, more or less, by the Carmagnols [poor/lower class people] of Arnheim. ... We learn, that the inhabitants of Neimegen are already weary of their French guests.[227]

Asgill was there for the whole of the retreat through Holland, after which he returned to London. He later recorded the time of his return as the winter of 1794,[228] but this is at odds with what *The Times* reported: "Yarmouth, March 7, Arrived, the *SWIFT*, from Embden. Robert Lawrie, Sir Charles Asgill and others came passengers in her."[229] And a history of the 1st Foot Guards, while providing more detail, further confused the timing:

The fleet sailed on the 24th of April [1795], and after encountering contrary winds and serious storms, made the coast of England on the 30th, but did not reach Greenwich till the 8th of May. Three companies of the First Guards disembarked at once, and were inspected by the king; the remainder of the battalion disembarked the next morning, and marched to the parade in St. James's Park. The two grenadier companies of the First Guards, under Lieutenant-Colonels Sir Charles Asgill, who had narrowly escaped the American scaffold, and Ludlow, the future Earl, together with the king's company under Lieutenant-Colonel Fitz-Gerald, arrived, on the 15th of May, at Greenwich, where they were met by the colonel of the regiment, the Duke of Gloucester, and on landing, found the king on the pier, who welcomed them back with much earnestness, and shook many of the private soldiers by the hand. They all received on their return eight days' leave of absence to visit their friends.[230]

Troubles in America Now Behind Him

On 26 February 1795 Asgill was promoted to the rank of colonel, at the age of nearly 33, and later that year commanded a battalion of the Guards at Warley Camp, 56 miles northeast of London, in Essex. Here the battalion trained for foreign service, but remained encamped for nearly three years.[231] Sophia undoubtedly joined him there, and did her best to be all that the commanding-officer's wife should be. How successfully, who can say. Their time in Essex was likely the best years they had together. The only blot on their happiness was the death of Sophia's mother, Hester Thomas, who passed away in 1796, leaving a grieving widower in Sir Chaloner.

Sir Charles and Lady Asgill were at the Warley Camp until January 1798 and soon set off for Ireland, a place they called home for the next 15 years.

CHAPTER FOURTEEN
Irish Uprising

On 1 January 1798, by now a major general and 35 years old, Asgill was appointed to the staff in Ireland, and was promoted Third Major of the 1st Foot Guards in November of that year.[232] The 1st Regiment of Foot Guards consisted of three battalions, each of which was nominally commanded by a major; the Third Major commanded the third battalion.[233] In a summary of his service he wrote:

> Appointed to the Staff of Ireland, was very actively employed against the Rebels during the Rebellion in 1798 and received the repeated thanks of the Commander of the Forces and the Government for my Conduct and Service. Remained on the Irish Staff til February 1802 when in consequence of the Peace I was removed and returned to England.[234]

Eventually, he spent 15 years of his army career serving in Ireland, then part of the Kingdom of Great Britain. But revolution was in the air. America and France had been through these traumas, and now the mood had spread to Ireland. The French Revolution provided inspiration to more radical members of the Volunteer movement in Ireland, who saw it as an example of the common people cooperating to remove a corrupt regime. Liam Chambers states that "the fundamental purpose of the United Irish rebellion of 1798 was the overthrow of the Irish administration based in Dublin; hence their primary military objective was the capture of the capital."[235] Asgill was back to just the place he had been during the American War of Independence. The populace were now, as then, divided by Loyalists, who wanted the country to remain under British control, and Rebels (known as "Croppies") who wanted to get rid of them.

It was a brutal but short-lived uprising, and Irish losses were high (10,000–50,000 Irish estimated combatant and civilian deaths, 3,500 French captured and 7 French ships captured, as opposed to 500–2,000 military deaths and approximately 1,000 Loyalist civilian deaths on the British side).[236] Once more, France came to the rescue of the rebels and, once more, Asgill served under Earl Cornwallis, who was the British Commander in Chief. Terrible atrocities were committed on both sides and far too many suffered, with little gained. Small fragments of the rebel armies of the summer of 1798 survived for a number of years and waged guerrilla warfare in several counties. It was not until the failure of Robert

Irish Uprising

Emmet's rebellion in 1803 that the last organised rebel forces, under Captain Michael Dwyer, capitulated. Small pockets of rebel resistance had also survived within Wexford, and the last rebel group under James Corcoran was not vanquished until February 1804. Religious, if not economic, discrimination against the Catholic majority was gradually abolished, but not before widespread mobilisation of the Catholic population under Daniel O'Connell. Discontent at grievances and resentment persisted but resistance to British rule now largely manifested itself along anti-taxation lines.[237] Asgill was there throughout these troubled times.

In May 1798, unrest broke out in County Carlow. General Asgill was given the task of restoring order there, and commanded forces that won victories over rebels at Baltinglass, Hacketstown and Carlow Town. In the latter place, according to a newspaper report, "four hundred of the misguided wretches were slain."[238] British regulars, supported by loyal Irish militia, forced rebels in the southeast of Ireland into County Wexford. There the rebellion in the region was snuffed out. Asgill called on a magistrate, Benjamin Bunbury of Killerig, to write a proclamation offering pardon to rebels who surrendered their arms and cease rebellious activities. The proclamation, posted on 29 June 1798, read:

> Whereas It is in the power of his Majesties Generals, and of the Forces under their Command, entirely to destroy all those who have risen in Rebellion against their Sovereign and his Laws, yet it is nevertheless the Wish of Government that those persons who, by traitorous machinations have been seduced or by Acts of Intimidation have been forced from their Allegiance should be received into his Majesties peace and pardon, Major General Charles Asgill commanding in the County of Carlow specially Authorised thereto, does hereby Invite all persons who may be now Assembled in any part of the said County against his Majesties peace to Surrender themselves and their arms, and to Desert the Leaders who have seduced them, and for the Acceptance of such Surrender and Submission the Space of Fourteen day's from the date hereby is allowed, and the Towns of Carlow, Leighlin Bridge, Gores Bridge, Borris, Myshall, Clonegal and Tullow, and something specified, at each of which places one of his Majesties Officers, and a Justice of the peace, will attend, and upon their entering their names, Acknowledging their Guilt, and promising good behaviour for the future, and taking the Oath

of Allegiance, and at the same time abjuring all other Ingagements contrary thereto, they will receive Certificates which will Intitle them to protection so long as the [sic] demean themselves as becomes good Subjects. And in order to render such acts of Submission easy and secure, It is the Generals pleasure that persons who are now with any portion of Rebels in Arms and Willing to Surrender themselves do send to him or to Lieutenant Col Mahon 9th Dragoons commanding at Carlow any number from each Body of Rebels not exceeding Ten with whom the General or Col Mahon will decide the manner in which they may repair to the above towns, so that no alarm maybe excited and no Injury to their persons to be offered.[239]

This proclamation was clear: "the Leaders who have seduced them" would receive no mercy, but those seduced would be afforded leniency. Asgill wrote to Lord Viscount Castleregh (the Irish Secretary during the Rebellion of 1798) to inform him of his success.

Kilkenny, 26 June 1798

My Lord, – Fearing the consequences that might result from allowing the Rebels who fled from Wexford to remain for any length of time in this country, I preferred attacking them with the troops I already had to waiting till a reinforcement arrived. My force amounted to eleven hundred men. The Rebels consisted of about five thousand. I attacked them this morning at six o'clock in their position on Kilconnell Hill, near Gore's Bridge, and soon defeated them. Their chief, called Murphy, a priest, and upwards of one thousand men were killed. Ten pieces of cannon, two swivels, their colours, and quantities of ammunition, arms, cattle, etc., were taken, and I have the pleasure to add that four soldiers who were made prisoners the day before, and doomed to suffer death, were fortunately released by our troops.

Our loss consisted of only seven men killed and wounded. The remainder of the Rebels were pursued into the County of Wexford, where they dispersed in different directions.

I feel particularly obliged to Major Mathews, of the Downshire Militia, who, at short notice and with great alacrity, marched with four hundred men of his regiment, and Captain Poole's, and the yeomanry corps of Maryborough, under the command of Captain Gore, to co-operate with me, Lord Loftus and Lieutenant-Colonel Rem, of the Wexford Militia, Lieutenant-Colonel Howard and Lieutenant-Colonel Redcliffe, of the Wicklow, Major Donaldson, of the 9th Dragoons, who commanded the cavalry, as well as all the officers and privates, are entitled to my thanks for their spirited exertions. Nor can I withhold the praise which is so justly due to all the yeomanry corps employed on the occasion; and I also beg leave to mention my aide-de-camp, Captain Ogle, Lieutenant Higgins, of the 9th Dragoons, who has acted as my brigade major. I have the honour to be, my Lord, your Lordship's most obedient servant, C. Asgill, Major-General.[240]

The Captain Ogle that Asgill wrote of was Sophia's 22 year old younger brother Thomas, born in 1776, an officer of the 58th Regiment of Foot who Asgill took on as part of his staff. Thomas was killed at Aboukir Bay, Egypt, in 1801.

Sophia had come to Ireland with her husband, and they found a lovely house in Kilkenny across the parade ground from the castle, but she was never happy there. She hated being separated from her large family, her friends in London, and the way of life they had so enjoyed in society there, when they were always finding reports in *The Times* of their engagements. She did not make friends in Ireland as easily as she had in London. She was always popular with men, but acquired the reputation of a coquette, and women were jealous of her. Their diaries are full of mean comments about her. Sophia was feisty, and could give as good as she got, but as a consequence of her unhappiness, she often tried to persuade her husband to allow her to return to London, which did not please him since he needed her to oversee his domestic arrangements.[241] Her sisters had homes in London, and they wrote often to tell her about their lives. It only served to make Sophia pine for her family and London even more.

Sophia was particularly close to her elder sister, Barbarina. Barbarina was a highly accomplished sculptress, artist and writer with friends who were very Bohemian. Sophia loved the world occupied by Barbarina, who was by now Mrs. Valentine Henry Wilmot. Barbarina was, in reality, Sophia's half-sister, since her mother (Hester Thomas, the daughter of a former Bishop of Winchester) had had an affair with her art teacher, Sawrey Gilpin, and Barbarina was the result of that love affair. It is not known whether Sophia's father, Admiral Sir Chaloner Ogle, knew this but he had brought her up as one of his own. However, Barbarina was well aware of her parentage, and proud that her biological father was one of the best painters of horses England has produced (his historical pictures were less successful; he was an animal painter only, and required the assistance of others to paint the landscapes and figures of his pictures). Barbarina took after him and herself became well known as a sculptress of horses. She corresponded with Gilpin, acknowledging their relationship and talking of their visits to one another. In one of his letters, written on 9 February 1800, Gilpin wrote to Barbarina: "Remember me kindly to Mr. Wilmot and my little granddaughter. God bless you, my dear child. Your very affectionate father. S. Gilpin."[242]

Barbarina Brand, Lady Dacre, née Ogle (27 May 1767 – 17 May 1854), Sister of Sophia, Lady Asgill. From the miniature, artist unknown, before 1908. In the possession of the Viscount Hampden, or his heirs. Wikimedia Commons.

Irish Uprising

Sophia was approached on a number of occasions by relatives trying to save their loved ones who had been leaders of the Croppies during the war. Their punishment for leading rebellious forces was the gallows, so it was not surprising that appeals came from relatives wishing to save their lives. They were often sisters of guilty men, and they came to Asgill's house in Kilkenny looking for mercy. Since he was so often not at home, Sophia greeted these women. One was Mrs. Fitzgerald, the sister of a rebel leader named William Farrell.

Farrell, in later years, wrote that he had been one who hoped that events would take a less violent turn than they had. Nonetheless, after he was captured he was sentenced to hang. His sister went on the same sort of quest to save his life as Dame Sarah Asgill had, sixteen years earlier, to save her son. Eventually Mrs. Fitzgerald decided to meet General Asgill personally, but when she went to their house in Kilkenny, she found Asgill away. Instead she met Sophia, who was waiting for his return. What ensued is best told in the words of Farrell himself:

> Lady Asgill received her with the greatest possible kindness, and on hearing her lamentable story, became so interested for her that she promised to use all her influence to serve her. In the course of some time the General returned and Lady Asgill herself opened the case to him in the most moving manner she possibly could, but he positively refused to interfere in it and said it was a case that did not come at all within his power, and that he could not on any account meddle with it and that the law should take its course. Lady Asgill, however, would not be put off, and continued to importune him in the most earnest manner, while he continued inflexible to all her entreaties and would not give way in the least, till at length she exclaimed, "Ah, General Asgill, you must not be too inexorable, particularly in the case of a boy, a young lad, quite a young lad, and you may recollect very well, when you were a young lad yourself, you were just in the very same predicament in America, and that it was a lady there saved your life, and upon my honour I'll save his life and you must do it."

In jail, Farrell was being prodded by an interrogator to implicate others who had participated in the rebellion:

> He came back immediately to let me know that if I gave information only against one, against one responsible man, I would be liberated; but he received the same answer and when all

efforts failed, Cornet Lowther came forward again and called me. As soon as I turned round to him, he opened up a large paper and read aloud the charges against me for which I was tried and sentenced to death – "but, your case having been laid before General Asgill," said he, "he has thought proper to change the sentence of death that stood against you to transportation for fourteen years." Here I bowed respectfully to him and everyone present expressed their approbation. "But," he continued, "you are so hardened a villain that I will go myself to General Asgill and make him withdraw his respite from you and I'll send you to hell to-morrow evening." The guard was then ordered to take me back to the guard-house and thus ended that evening's tragedy.

The interrogator's threats proved hollow. Farrell concluded his account of his ordeal:

> It will no doubt seem strange to any person living in the present times how any prisoner could be treated in the manner I was; I underwent four most extraordinary changes in one day. I was under sentence of death in the morning; I received General Asgill's respite from death in the evening and afterwards was brought to the place of execution, where sentence of death was again passed on me; and after standing in that perilous state as long as they pleased, was respited again … the whole aim of it was to compel me to give information.[243]

The whole episode, for William Farrell, had run along the same lines as Asgill's own experience in 1782. Rumours of "thrice to the gallows" ran through his mind once more. Farrell's sister had played the part of Dame Sarah, and Sophia had played the part of the late lamented Queen Marie Antoinette. Ironically, at the end of the day, Asgill played the part of the Continental Congress, and spared a man's life. Although he had insisted that a pardon was not within his remit, he eventually allowed himself to be swayed, and ordered that Farrell's life be spared.

Lady Sophia had deftly demonstrated an attribute that she later related to her friend Lady Melbourne, who wrote: "What Ly Asgill says is ye real test of a Womans cleverness – that is managing her Husband – she says it is what every Woman ought to Study for her own happiness & her Husbands too – & she never can think any one can have any pretentions to cleverness who does not do so, by some means or other."[244]

Irish Uprising

This was not the only such experience for Sophia. A woman named Mary Maher had two brothers, William and James, who were sentenced to be executed. Mary carried a pleading letter to the home of "the singular and eccentric, but not unamiable wife of General Asgill, requesting her to intercede with her husband to save her brothers.":

> Mary Maher delivered this letter to Lady Asgill's door, but was barred from entry by the footman. Found later by a visiting dinner guest, Colonel Butler, Mary had fainted on the doorstep still clutching the precious letter. Lady Asgill was informed and immediately took the poor girl in and ensured she was administered aid. When Mary recovered sufficiently to recount her tale, Lady Asgill entreated her husband's pardon for these two Maher brothers. By her wily tactics, it was granted and Mary hastened to the prison ship where her brothers were being held, now clutching the precious 'pardon'. Poor Mary Maher arrived with this missive too late – her two brothers died of a fever ravaging the prisoners on board. They were buried that same night and Mary was found next morning trying to exhume their bodies to bury them in their home town. She had lost her mind – her travails in trying to save their lives had taken the saddest toll of all and Mary lived out the rest of her life in a state of lost sanity.[245]

Regardless of the tragic outcome, once more Sophia had shown what a persuasive woman she could be, and Asgill had again shown compassion.

The redeeming feature of these ghastly episodes is that in many areas there were officers and gentlemen who did not accept the need for severity. General John Moore avoided holding any courts martial at all. General Ralph Dundas in Kildare remitted most of the death sentences to transportation. At Birr, Sir Laurence Parsons wrote to protest against the flogging to death of a man by two magistrates. And in a large part of Connaught the gentry, according to Lord Altamont, "disapproved of the measure entirely and having any one opinion upon it will not cooperate in the execution of it." Yet for every loyalist in the threatened areas of Carlow, Kilkenny and Queen's County who disapproved of severity, there were a hundred who felt that the policy was lenient to a fault. Asgill would not let the majority opinion hold sway, and resisted excessively harsh measures. "This country would have been in the same situation with Wexford" declared one loyalist, but for Sir Charles Asgill's "activity and exertion." It was proposed to present him with a ceremonial sword. The city of Kilkenny presented him with a snuff box for his "energy and exertion" which was praised by the Loyalists; it was inscribed, "Cando Asgill/Bart etc etc/ *Municipium de St. Canice/ Gratum Memor/ D.D & Pax Interna/ Conservatu Rebelles/ ad Kilconnel/ debellati.*"[246] The loyal people of Clonmel, Tipperary, presented him with a silver hot-water urn in 1801. The inscription on the urn reads: "PRESENTED by the Inhabitants of the Town and Neighbourhood of CLONMEL to MAJ^r. GEN^l. SIR CHA^s ASGILL BAR^t. in token of their great regard for His unremitting exertions as General Commanding in the district in defeating the Schemes of the Seditious and Protecting the loyal Inhabitants. CLONMEL MDCCCI."[247] In the words of one historian, he was "humanity personified."[248]

The hot water urn presented to Charles Asgill by the people of Clonmel in honour of his defeat of Irish rebels in that town in 1801. By courtesy of the late Professor William Bensen.

Irish Uprising

Not all Irishmen remembered him well, though. Bitter fighting had taken place, and atrocities had occurred on both sides. A song, *Sliabh na mBan,* came out of the rebel camp, and they say it was dedicated to Asgill. It is still played today, and is a haunting, but very pleasant melody.[249]

CHAPTER FIFTEEN

On the General Staff in Dublin

After the 1798 rebellion in Ireland was suppressed, the Asgills had a brief sojourn in England. On 9 May 1800 Asgill, now 38, left the 1st Foot Guards to become Colonel Commandant of the 2nd Battalion of the 46th (South Devonshire) Regiment of Foot.[250] But the 2nd Battalion had been raised only for wartime and was disbanded in 1802. With the battalion no longer in existence, he was put onto half-pay, a sort of reserve status that allowed him to return to active service if needed.[251] He and his wife finally had a chance to be back amongst family and friends in England, and Sophia was very happy to be home. Asgill was relieved to see her so relaxed, but knew it wouldn't be for long.

In 1801, during their time in England, he found himself defending the right of Henry Ellis (who lived in the neighbourhood of Kilkenny) to be properly remunerated for invaluable intelligence he had provided during the rebellion. His information had made a significant contribution to the suppression of the rebels, but he paid a severe price for his loyalty after the fighting was over. His neighbours persecuted him, tried to kill him; and ruined his business as a miller. The British government was very slow indeed to pay his annuity of £30 per annum for life, and he became a ruined man. Lord Castlereagh and Asgill took up his cause to see that he was properly compensated. Nevertheless, Ellis suffered greatly because of "his grievious situation, and losses, on account of his loyalty to his king and government." When he died he was buried in "an inverted burial in the unconsecrated ground of his own farmland ... The lack of any memorial stone or grave marker is mysterious. Perhaps one was destroyed generations ago by inhabitants of the locality or never erected in anticipation of such an act."[252] Being buried inverted – vertically with head down and feet up – was a ritual reserved for criminals and miscreants. There were many others in the same situation, and, as with the American War, there was much suffering amongst the loyalist population.

Their time back in England did Sophia's health the world of good. She had been yearning for her family and her social life; her health was not always good and she desperately needed to get away.

Her younger brother Charles, a naval officer returning from duty abroad, had announced his engagement to Charlotte Gage, the daughter of General Thomas Gage, who had been Britain's Commander-in-Chief in the early days of the American War of Independence. Sophia's older sister, Arabella, wrote to a friend of the family, General Thomas Graham, "I think you will be pleased with her." She also mentioned that Richard Streatfield, who had been a widower since the 1796 death of Jane Ogle, Sophia's elder sister, had now remarried.[253] Sophia was keen to meet these new members of her family.

By this time Barbarina and Valentine Wilmot had separated and her purse-strings were tightly bound, especially having a daughter to bring up on her own. The sisters had also been concerned about one of their father's butlers, who had got into financial difficulties. They were all fond of him, so they wanted to help him and his wife get through their troubles. General Graham had been generous to a fault in all the gifts he heaped on the Ogle sisters, so Barbarina hoped Graham would give £270 to save the butler from financial ruin. Sophia and Arabella were not in a financial position to help, but Barbarina proposed to sell some of her drawings to meet her share. After Graham *had* helped out, Barbarina wrote to say: "I could not help telling them your wonderful goodness — don't scold — my conscience and fifty other things insisted on it."[254]

Because Asgill was on half pay during their time in London, waiting to be reappointed to other duties, their financial position was not good, so Sophia and he could not chip in. They did what they could. They planned a holiday in Margate, and invited Barbarina to join them. She complained, saying she would much prefer the Sussex coast. "Margate is resolved on," she wrote to tell Graham. "It quite provokes me. I am told the bathing is the very worst in the world, & the odiousness & vulgarity of the place intolerable … The journey alone will half ruin me."[255]

The caustic Harriet Cavendish wrote to say that their party rode daily to Ramsgate, and that her coachman had commented on Sophia's unladylike behaviour: "he is much shocked at her showing her legs, which, to be sure, she does freely." Continuing with her bile, she wrote of how she disapproved of the girls wearing riding habits on a family visit to see their aged father in Worthy. She wrote, "here comes Sophia in a scarlet riding habit." Barbarina wished that Graham had been with them too, so that they could all "be brothers and sisters in perfect harmony and perfect health together." She assured him that Sophia was in her element and that she was "charming, cheerful and entertaining and that Sir Charles is beaming with satisfaction."[256]

The Charles Asgill Affair

On another occasion, Henrietta, Countess of Bessborough, when writing to Granville Leveson Gower, described a situation which arose when the Duke of Richmond had been driving his housekeeper in a wagon, which overturned. She said that it was fortunate that he "fell onto her ample bosom, which cushioned his fall". But she couldn't resist a dig at Sophia, saying that had she been his travelling companion, he would have fallen onto "a sharp Angle, little better than the road itself."[257]

Henrietta Bessborough had more to say about Sophia:

> We do not see a great deal of company, only the Melbournes (they have but one Son here, George) and Ly. Asgill and Mrs. Wilmot, who, tho' they live at Margate, come very often. I like the latter extremely. I do not know what to make of the former. Sometimes I am inclin'd to like her partly from her coaxing manners, and because I have great respect for your taste, but every now and then she puts me quite out of patience with her coquetry and affectation, and seems to me quite a fool. This is when I see her twist herself into ten thousand shapes, affect childishness and naivete, and take as much pains to turn the heads and secure the Hearts of Ld. Colleraine and Mr. Graham as she would to attach you or the most delightful creature that ever liv'd. I must tell you I am writing in a bad moment, because she provok'd me first by Playing twenty tricks this morning, and when I determin'd to get over that, and was beginning to like her again, she ask'd me very seriously whether I did not think Adair a man of extraordinary talents and great eloquence and capacity. I look'd up to see if she was in Earnest, and she really was.[258]

Later that year they were back in the Isle of Thanet as they had been invited to dine in Broadstairs. The event was reported in the newspapers: "On Tuesday Lord Say and Sele[259] gave a Dinner at Broadstairs, and in the evening a concert of vocal and instrumental music. The Duchess of Devonshire, the Duchess of Manchester, Lord and Lady Bessborough, Lord and Lady Carysfort, Sir Charles and Lady Asgill, Lord and Lady Melbourne, Mr W. Lambe, the two Miss Lambes, etc. etc. were of the party."[260]

On 18 March 1803 Asgill was reappointed to the army staff in Ireland, and placed in the command of the Eastern District, which included the Garrison of Dublin. He was in command during the rebellion which broke out in the city on 23 July 1803.[261] It was led by Robert Emmet, whose plan was to seize a few strategic positions within the city and then wait for others to rebel. Their main target was Dublin Castle, which was reported to be lightly guarded and was a highly symbolic target as the seat of English/British government in Ireland. When the uprising failed, Emmet fled into hiding but was captured on 25 August. On 20 September he was executed in Thomas Street, hanged and then beheaded once dead.[262]

    Asgill was promoted to Lieutenant General in January 1805 at the age of 42.[263] Late that year, a soldier named Patrick Reardon of the 2nd Battalion, 28th Regiment of Foot sent a petition to him complaining about something going on within the regiment; the specific complaint has not been found, but it appears to have concerned the ways that discipline was being enforced in the regiment by the commanding officer, giving Reardon no reasonable recourse but to bring the matter to a higher authority, namely General Asgill. Asgill, ever concerned with the welfare of soldiers, took the petition into consideration.[264]

    The commander of Reardon's battalion, Lieutenant Colonel Charles Philip Belson, learned of the petition and took immediate punitive action. He ordered that Reardon be tried by a regimental court martial for writing the petition, apparently considering it mutinous or insubordinate. The court was assembled hastily and the trial occurred at night, suggesting an intention to punish Reardon before General Asgill investigated his allegations. Reardon was found guilty and sentenced to immediately receive two hundred lashes (which was not particularly severe for the era). After Reardon had recovered from the wounds inflicted by this punishment, Lieutenant Colonel Belson demanded that he be forced to spend "two or three hours a day at the dumb-bells" in spite of Reardon's known "infirmity and inability" to bear fatigue.[265]

    When Asgill learned of this, he ordered Belson to be tried by a general court martial on six charges: disobedience of orders (because he had tried Reardon while Asgill was still considering his petition); disrespect towards Asgill in a letter Belson wrote concerning the case; cruelty in trying Reardon by a regimental court martial when his petition was being considered; improperly conducting the court at a late hour and ordering the punishment to be immediate; unmilitary and "unofficerlike" behaviour in continuing to punish Reardon after the lashings; and cruel conduct by ordering sergeants to more frequently strike soldiers with their canes.[266]

The Charles Asgill Affair

The trial commenced on 30 December and took several days to complete. Among other things, the court determined that Belson did not know that Reardon's petition was being considered by General Asgill, so could not be held accountable for bringing him to trial. Belson was acquitted of all charges except irregularly conducting Reardon's trial, for which the court sentenced only a reprimand. Knowing that a public reprimand was a significant humiliation for an officer, the court made the further recommendation that the King grant clemency to Belson – that is, allow the sentence to not be carried out. The members of the court were "satisfied he was actuated by no improper intention, or motive of disrespect, to Lieutenant General Sir Charles Asgill."[267]

The King, in his role as commander in chief of the army, reviewed the trial proceedings and disagreed with the recommendation of the court. He deemed it "to be indispensably necessary for the support of discipline, that the sentence of reprimand, which in itself appears to be sufficiently lenient, should be confirmed." The King also noted that, although he acquiesced to acquittal on the charge that a regimental court martial should not have been called when the soldier's petition was under consideration by Asgill, it was apparent to the King that Belson did know that the petition was under consideration. Asgill's own opinion on the proceedings is not recorded, but the King's response certainly vindicated his bringing Belson to trial. It is one more testimony to Asgill's sense of fairness and compassion for the soldiers he commanded.[268]

Asgill was appointed Colonel of the 5th West India Regiment in February 1806,[269] Colonel of the 85th Regiment of Foot in October 1806,[270] and Colonel of the 11th (North Devonshire) Regiment on 25 February 1807.[271]

The 11th Regiment had just returned home from Barbados where they had been posted for six years. When they were inspected on 14 January 1806 they had only about a third of their normal strength. The inspecting officer found them "worn down by the climate and unfit for active service."[272] After six years in the harsh West Indian climate their condition was not to be wondered at. This was the state of the regiment when the colonelcy passed to Asgill. He, a down-to-earth, non-political soldier, was eminently qualified to rebuild and safeguard the interests of a solid regiment of the line. Under his aegis the 11th, in company with the rest of the British Army, would lift itself out of what one historian called, with hyperbole, "the fifty years' pit of defeat and neglect into which it had fallen."[273]

On the General Staff in Dublin

On 28 June 1808 the Duke of York, commander in chief at the Horse Guards, had this to say: "When Lieutenant General Sir Charles Asgill was appointed to the command of the 11th Regiment in February 1807, the effective strength of that corps was only 162 rank and file, but which by his exertions had now been completed to its present establishment." At its full strength of 936 men, the Duke wrote that "recruiting must necessarily cease" unless a larger established strength was authorized. Not wanting to put an end to successful recruiting, he recommended adding a second battalion to the regiment, effectively doubling its size.[274]

Asgill enjoyed Ireland, but Sophia missed her family badly and was often not in the best of health. She made the most of it though, and was in the forefront of Dublin society:

> Lady Asgill gave a grand route on Thursday evening last at her house in Merrion-square, at which most of the nobility and gentry in town were assembled; his Grace the Lord Lieutanant honoured the party with his presence, but her Grace the Duchess of Richmond, who had only arrived from England the preceding day, did not make one of that elegant and fashionable party.[275]

Asgill's nickname of "little Charlie" and his strict sense of impeccable sartorial elegance were well known in Ireland, as the following anecdote attests:

> "I say, Dudley, who have we at dinner to day?"
>
> "Harrington and the Asgills, and that set," replied he, with an insolent shrug of his shoulder.
>
> "More of it, by Jove," said O'Grady, biting his lip. "One must be as particular before these people as a young sub. at a regimental mess. There's not a button of your coat, not a loop of your aiguilette, not a twist of your sword knot, little Charley won't note down; and as there is no orderly-book in the drawing-room, he will whisper to his grace before coffee."[276]

After spending some time in the discharge of his parliamentary duties, and in attending to the improvement of his estates in Scotland, in 1807 the Asgills had the great pleasure to see the beloved friend of the Ogle family, Colonel Thomas Graham, serving on the staff in Dublin. This pleased Sophia immensely. But Graham's letters home showed that he was not happy; he hated so much travelling in chaises on dreadful roads, and

The Charles Asgill Affair

he was not happy with his horse, either, which was far too large and beat itself up on the bad roads. His cousin, Robert Graham, tried to ease his suffering by sending cognac and a hamper with the best marmalade. Thomas Graham was also very concerned about Sophia's health, requesting Robert find two pairs of the best Shetland stockings for her, in a small man's size since ladies stockings were too short for her – she was tall and willowy in stature and had always had difficulty finding hose to fit. She was always grateful for the kind gestures made by Graham to herself, and her Ogle sisters, who frequently received lavish gifts from Graham. He also wanted a little Arabian mare by the name of Tilburnia to be sent to Asgill, since he was unable to find anything like her in Dublin.[277]

Always concerned for Sophia's welfare, Graham told her that he was worried about one of her dogs, which had gone mad (presumably rabies) and he warned her that a woman had died from being bitten by a mad dog. Sophia refused to do anything more than keep the dog in the stable, unchained, rather than in the house.[278] From childhood, she had always wanted her dogs close to her.

So worried was he about Sophia's health, cousin Robert was asked to find a boat to get her away from Ireland. This caused some embarrassment, because Sophia had not impressed on her husband her wish to leave. It seems she wanted to visit her sisters, but Asgill would have been unable to manage his household in Ireland without her. It turned out that Graham and Sophia had hatched a secret plan. Sophia had caught a cold, although she never left the house, and this had worried everyone as she had been threatened with a chest infection. Graham had explained to his cousin Robert, that he was especially worried on Sophia's account, and so "with the help of her physician's opinion to try to bring this favorite & indeed necessary object to bear," they thought that if the plans were all in place, Asgill would agree. It certainly threw light on the extent to which Sophia's health mattered to so many. Asgill took the opportunity to get Sophia away from a climate which didn't suit her, in the summer of 1807. After a pleasant interlude with the Duke of Devonshire, at Hardwick, they all had a most enjoyable couple of weeks by the seaside, in Scarborough.[279]

To cheer herself up, Sophia never missed the chance to entertain, and one of her dinners was recorded by one of the guests:

> Having to plead as an excuse to Sir Charles and Lady Asgill, for a late attendance at their dinner-party, that if I were served up with the game of the second course, it would be because I was first to

assist at a tea-party given to Mr. Kirwan, they expressed not only surprise at his residence in Dublin, but an anxious curiosity to be of the tea-party. Not to take the philosopher by surprise, the proposition was made to him by Lady C and myself — and I remember his answer was, "Madam, I am always pleased to mingle with people of the world. I never knew one, even the lightest and most frivolous, from whom something was not to be learned, that threw a light upon the follies and virtues of society. I once lived much in the world of fashion myself: and was as foolish and as vain as the worst. But I stipulate for my own hours, my own *tay,* and my own *tay-pot."* — This being agreed, the party assembled in Lady C.'s drawing-room, at the usual fashionable hour for morning visits. Under the pretence of bringing his staff, Sir Charles Asgill was accompanied by his amiable nephew and A.D.C. Captain Bouverie, [the son of Sophia's sister, Arabella] and several other young officers: and Lady Asgill smuggled in General and Lady Augusta Leith. In short, the whole 8 o'clock dinner party of Merrion-square were sealed at my sister's tea-table before six.[280]

The guest went on to note,

there was a previous inclination on the part of the fashionables towards mystification; and that a very active system of quizzing had been organized by the two *grandes dames de par le reconde* in which every beau present was to have played his part – (within, however, be it acknowledged, the bounds of perfect good-breeding, — a virtue never transgressed with impunity in the society of the polished and courteous Sir Charles Asgill).[281]

Having established a second battalion of the 11th Regiment of Foot, Asgill had to pay to equip his men out of his own pocket; he then experienced difficulty receiving a rebate from the Treasury. So, Lieutenant General Sir Arthur Wellesley (later the Duke of Wellington), the Chief Secretary to the Lord Lieutenant of Ireland, wrote from Dublin Castle to his brother Henry, Joint Secretary to the Treasury, on 3 January 1809:

I enclose some papers which I have received from Sir Charles Asgill relative to the issue of 8 months off reckonings for the Second Battallion 11th Regiment. He is entitled to this issue, but the ground on which he desires to have the money at an early period is that he raised the regiment in this country and purchased for them the accoutrements and other articles the expense of which this advance is intended to defray and that whenever articles of this description are purchased in Ireland they must be paid for in ready money, as the tradesmen are unable to give credit as they do in England. An officer, therefore, who incurs the expenses here ought to receive the money from the Treasury to defray them as soon as possible, otherwise he must borrow the amount which is his due and pay interest for it. I shall be obliged to you if you will endeavour to manage this matter in such a manner as that Sir Charles Asgill may receive his 8 months off reckonings immediately.[282]

Asgill must have been grateful for Sir Arthur's intervention, because his financial situation had become strained once more, since he had incurred the costs of money-lenders; his reserves were never as substantial as they should have been, had he not indulged in gambling his fortune away in his youth.

CHAPTER SIXTEEN
Jemima Sophia (Ogle) Asgill

Sophia pronounced her name so-FYE-a, perhaps to lend assonance to her first two names, Jemima Sophia. She was "the singular and eccentric, but not unamiable wife of General Asgill."[283] She was all of those things. Beautiful and irresistible to men, she could charm any man and get whatever she wanted from them, but she was criticised by women. This most likely stemmed from jealousy of her beauty and her attraction to the opposite sex. It was part of the reason she had not settled so well in Ireland.

Her entire family were eccentric. As children, their father would be driving his carriage and fail to notice if one of his many offspring fell off![284] And her mother, Hester Thomas, whose father was once the Bishop of Winchester, was also charming, but equally eccentric. Hester had charmed Sawrey Gilpin, her art teacher, enough to bear his child, Sophia's half-sister Barbarina. None of this stopped the Ogle family from being a loving and close-knit unit who would always help one another out. They were originally of Northumberland stock, going back to the fourteenth century.[285]

Sophia got to know the novelist Maria Edgeworth in Dublin. Edgeworth's 1806 novel *Leonora* features a character named Lady Olivia who was rumoured to have been based on Sophia, portraying her as a "coquette."[286] Edgeworth wrote a letter to her aunt dated 3 December 1809 that explains more about Sophia:[287]

> [Miss Whyte] told us a great many good anecdotes of Lady Asgill - of whom she has seen a great deal, and it was for some time difficult for us to determine whether she was her friend or her enemy but at last this point was determined by her account of a battle royal between these two belles at Miss Whyte's own table lately in Dublin. Lady Asgill began the attack thus "Miss Whyte do you know the good people of Dublin are beginning to abuse you quite as much as they abuse me". "Oh no, I hope not quite so bad as that" – quoth Miss W. "Why though they abuse me, I'm certainly very popular" reasoned Lady A – "for if I invite 60 people to my dinners or my concerts not one of the 60 send an excuse. They all come to my parties". "Oh that is no proof of popularity" replied Miss W "for your ladyship knows that if one came down from the gibbet and gave good dinners and good music

The Charles Asgill Affair

they might be sure of having everybody at their parties." The conversation went on from popularity to notoriety - then the word famous was brought in by some of the company and a Mrs Parkhurst (the English lady who brought in the message about comedy from Sheridan) brought in the word infamous. I don't exactly know how but Lady Asgill, who has, it is said, infinite command of temper, coolly in her high keyed voice "Does Mrs Parkhurst mean to say that Miss Whyte and I are infamous?"

Since the Ogle family were such close friends of Thomas Graham they were often in one another's company, especially in London. Once Graham had become a widower, in the spring of 1792 when he tragically lost his wife, Mary Cathcart, he immersed himself in a military career. He was a Scottish aristocrat, politician and British Army officer who became a hero during the French Revolutionary Wars and the Napoleonic Wars.[288] Sophia and her sisters loved him as a brother, and he them. With their army careers in common, Asgill got on well with him too. Once the Asgills were back in England, in December 1812, Sophia was invited to Woburn Abbey (the family seat of the Russell family, the Dukes of Bedford), and Graham was there too.

One of the guests at Woburn Abbey, Lady Frances Shelley (the daughter and heiress of Thomas Winckley of Brockholes, Lancashire; a noted diarist, and close friend of the Duke of Wellington[289]), wrote to Lady Spencer (the eldest daughter of the Whig politician, Sir George Spencer, 2nd Earl Spencer (1758–1834)) after one of her visits to Althorp:

Woburn December 1812.

Dear Lady Spencer, ... As soon as we left the dining-room, the Duchess [of Bedford] went to her nursing employment (after a little edifying conversation on the subject) and we dispersed into different parties, through an enfilade of six rooms. The gentlemen soon joined us, and in the first, Shelley [her husband, Sir John Shelley] got a companion at billiards. In the next, Lady Asgill established herself in an attitude, lying on the sofa with Sir Thomas Graham at her feet. In the next, a sober rubber at whist. In the next, Lady Jane and Miss Russell at a harp and pianoforte (both out of tune), playing 'The Creation'! Alas! It was chaos still! And, in the long gallery, a few pairs were dispersed on the sofas; others sauntered from room to room. I joined the latter, and talked of furniture, china, and ormolu, till the subject was exhausted. I was bored to death, and *triste à mourir;* the *tête-ê-tàte* forming a barrier to the billiard-room! At last I established myself at a writing-table in the card-room. Scarcely was I seated, when the Duchess entered; and, collecting her romping force, of girls and young men, they all seized cushions, and began pelting the whist players. They defended themselves by throwing the cards and candles at her head; but the Duchess succeeded in overthrowing the table, and a regular battle ensued, with cushions, oranges, and apples. The romp was at last ended by Lady Jane being nearly blinded by an apple that hit her in the eye! Shelley, before that, had been almost smothered by the female romps getting him on the ground, and pummelling him with cushions.

To this succeeded 'Blind Man's Buff.' Triste, disgusted, and cross, in spite of my good resolutions to bear any amount of folly (dear Lady Sarah, forgive me), I stole off to bed. As I passed along the corridor, I almost expected that the picture of Lady Rachel Russell would start from its frame, at seeing her favourite residence turned into a *guinguette*. But the picture, and the representative of the House of Russell are equally accustomed to, and unaffected by, such scenes; and living, as the Duke does, in the languor created by the dearth of intellectual amusement, can you wonder that he should, in despair, try to enjoy the physical distraction even of 'Blind Man's Buff'? Thus far is for the Nest. But what their pure minds would think impossible is the disgusting familiarity of Lady Asgill and Sir Thomas Graham, who, though in the field a hero, is in love a dotard. To give you a specimen — Lady Asgill yesterday said to me, in speaking of the house at Woburn: 'We have the apartments next yours. They all communicate, which is extremely comfortable. Sir Thomas's is next yours. I have the next, and my sister, Mrs. Wilmot [Barbarina Ogle], the third.

You have seen too much of the world to be surprised at anything, but to me this parade was both new and disgusting[290]

Had they been lovers that night, or was this female tittle-tattle? In writing about this episode, Graham's biographer says that Lady Shelley "suggests an illicit relationship." He goes on to say that perhaps she would know, given she was the "long-suffering, patient wife of an unfaithful husband."[291]

Graham and Sophia had, for many years, maintained a secret correspondence with each other. These letters commenced in 1795, when his regiment had returned from the Ile de Noirmoutier, an island off the Atlantic coast of France, and continued for the rest of her life.[292] The two had agreed to destroy each other's correspondence, but some letters survived. Many of Graham's letters to Sophia were written from the battlefield, during his military campaigns in Europe.[293] He kept only one of her letters though, presumably written before they agreed to destroy their correspondence.[294] Sophia, on the other hand, kept all of his (those written until their Woburn Abbey encounter, but none written after). A graphite drawing of Sophia's setter dog was created at Graham's family home, the Lynedoch Estate in Scotland, by Charles Loraine Smith.[295] Had that been a secret tryst as well?

One of Sophia's friends was the poet Mary Tighe. Between 1806 and 1809 she wrote "The Shawl's Petition, to Lady Asgill," described as a friendship poem.[296] In it, a shawl from India that Sophia gave to Tighe writes a petition to its former owner, expressing its unhappiness at having been exiled as a gift, and its hope to return, at least for an occasional visit.

> OH, fairer than the fairest forms
> Which the bright sun of Persia warms,
> Though nymphs of Cashmire lead the dance
> With pliant grace, and beamy glance;
> And forms of beauty ever play
> Around the bowers of Moselay;
> Fairest! thine ear indulgent lend,
> And to thy suppliant Shawl attend!

> If, well content, I left for thee
> Those bowers beyond the Indian sea,
> And native, fragrant fields of rose
> Exchanged for Hyperborean snows;
> If, from those vales of soft perfume,
> Pride of Tibet's far boasted loom,
> I came, well pleased, thy form to deck,
> And, from thy bending polished neck
> Around thy graceful shoulders flung,
> With many an untaught beauty clung,
> Or added to thy brilliant zone
> A charm that Venus well might own,

Jemima Sophia (Ogle) Asgill

Or, fondly twined, in many a fold
To shield those lovely limbs from cold,
Fairest! thine ear indulgent lend,
And to thy suppliant Shawl attend.

Oh! by those all attractive charms
Thy slender foot, thine ivory arms;
By the quick glances of thine eyes,
By all that I have seen thee prize;
Oh! doom me not in dark disgrace,
An exile from Sophia's face,
To waste my elegance of bloom
In sick and melancholy gloom;
Condemned no more in Beauty's train
To hear the viol's sprightly strain,
Or woo the amorous zephyr's play
Beneath the sunbeam's vernal ray;
Banished alike from pleasure's scene,
And lovely nature's charms serene,
Oh, fairest! doom me not to know
How hard it is from thee to go!

But if my humble suit be vain,
If destined to attend on pain,
My joyless days in one dull round,
To one eternal sopha bound,
Shut from the breath of heaven most pure,
Must pass in solitude obscure;
At least to cheat these weary hours
Appear with all thy gladdening powers,
Restore thy sweet society,
And bless at once thy friend and me.
Oh, fairer than the fairest forms
Which the bright sun of Persia warms,
Though nymphs of Cashmire lead the dance
With pliant grace, and beamy glance;
And forms of beauty ever play
Around the bowers of Moselay;
Fairest! thine ear indulgent lend,

The Charles Asgill Affair

And to thy suppliant Shawl attend![297]

Was that sorrowful shawl ever returned to Asgill's beautiful wife? This, we don't know.

Sophia was in her element at parties; while enjoying the company of men more than women, she was never short of a witty repartee. She loved her husband though. For his part, he loved her deeply and couldn't live without her.

"Lady Asgill's Setter – taken at Lynedock 1809." Graphite drawing by Charles Loraine Smith. Sold by Somerset & Wood to Anne Ammundsen in 2019.

CHAPTER SEVENTEEN
Retirement from the Army

Sitting at his desk in Dublin Castle one day, General Asgill was shocked to receive a letter from the Duke of York, written on 3 January 1812, telling him that on account of Lieutenant General Sir John Hope's appointment to command of the forces in Ireland, "you will unavoidably be discontinued on the staff of the Army." The Duke went on to say:

> At the same time that I was honoured with the Prince Regent's Commands, upon this subject, I received His Royal Highness' Express Directions, to signify to you, His entire approbation of your zeal and ability, in the Discharge of the Duty which has been entrusted to you, and it is with great satisfaction, that I embrace this opportunity of assuring you, of my best thanks, for the able assistance I have invariably experienced from you, and of my regret that the Military Arrangements do not admit, for your employment upon the Staff of the Army, in Great Britain, at present.[298]

Hope (4th Earl of Hopetoun) was appointed Commander-in-Chief, Ireland, and was admitted to the Irish Privy council in 1812.[299] He went on to command the First Division under The Duke of Wellington at the Battle of Nivelle and at the Battle of the Nive in 1813.[300] His time in Dublin was very short-lived, compared to Asgill's 15 year service in Ireland. Asgill was almost 50 years old at the time, and explained, in his reply to Colonel John McMahon, Private Secretary to the Prince Regent:

> I shall for the first time in my life return to England with a reduced income, and without any employment, which is not very pleasant to my feelings after an uninterrupted service of thirty four years, fifteen of which have been spent on the Staff of Ireland. ... As it is probable, the Prince Regent's Establishment will be soon arranged, I beg leave to mention to you that I should consider myself highly honoured if His Royal Highness would be pleased to appoint me to any Situation he might deem me competent to fill ... I should esteem it a very great favour if you will have the goodness to take an opportunity of making my wishes known.[301]

Asgill (apparently) never heard back from McMahon, and handed over command of the Staff in 1812.[302] The following year Lord Wellington wrote to Lord Bathurst concerning the financial circumstances of British General officers, saying:

> The General officers of the British army are altogether very badly paid, and, adverting to the deductions from their pay, they receive less than they did fifty years ago, while their expenses are more than doubled[303]

Entry to the State Apartments, Dublin Castle. Much of Asgill's military career was spent at this Irish post. Wikimedia Commons.

When the time came to leave, Asgill realised just how sad he was to depart Ireland, although that could not be said of Sophia. Thanks were expressed to him, which moved him greatly:

> The Sheriffs severally expressed their thanks on their healths being drank, with manly and becoming sensibility, like men who knew the value of honest fame, and who are resolved to preserve it by independence, purity and integrity. Sir Charles Asgill was very happy in taking the opportunity, while returning his thanks, to take his leave also of this City, the garrison of which he had

Retirement from the Army

commanded for nine years. He observed upon the obligations which he owed the national hospitality of this country, the happy understanding which always prevailed between the soldiery and the loyal Citizens of Dublin, and he declared that the sentiments he then expressed, should be for ever cherished in his heart, and that he should never cease to wish for the happiness and prosperity of Ireland. — Sir Charles spoke with great effect an honest effusion of feeling, which was acknowledged amid applauses, and the kindest emotions.[304]

So ended his 34 year career as His Majesty's loyal servant. He did not feel too old to continue, and very much wished that he had had the opportunity to do so. He was also concerned about his financial situation, with reduced payments from the army to help with his costs. His Colonelcy of several regiments continued, though, which pleased him.

Once they were back in London, Sophia's sister, Arabella, wrote a letter to Thomas Graham to let him know that Barbarina was "bustling about London and reassured him over Sophia's health and good looks," going on to say:

> she endeavours to make the best of everything. Now and then she declares she detests being obliged to moderate her expenses — however, she is resolved to be prudent — in many respects I think she is much altered, and certainly looks very differently on the gayeties of a London life ... We are all poor as Church Mice and provisions beyond measure dear[305]

She continued by explaining that the weather was appalling, and colder than she had ever known. Were Graham with them they would have made great fires and fed him their finest beef, although she knew he preferred puddings!

Charles and Sophia went back to 6 York Street[306] and settled down to enjoy their retirement years. The street is now named Duke of York Street, and the house has since been demolished, it is believed in the late 1960s. It was a lovely house and it didn't take long for them to get back into London society, enjoying the things they always had. Sophia was elated to be home, permanently. They resumed their old friendships, and were often in Bath, enjoying the waters there, or Brighton, both of which they had visited many times previously. In 1810, three Guards officers, Captain Rees Howell Gronow, Captain Jack Talbot and Colonel Dan McKinnon established the Guards Club, at the coffee house located at 88

The Charles Asgill Affair

St. James's Street; it is highly likely that Asgill joined on his return from Dublin.[307] It was just what he needed, and would have enabled him to meet up with old friends whenever the opportunity arose.[308] He was also a member of the Freemasons Lodge No. 537, which met at the Star and Garter public house in Pall Mall. This was a private lodge for the Duke of York (the Master of the Lodge), who didn't see the need to furnish the Grand Lodge of England with its membership list or even pay any fees to the Grand Lodge for these members! One of the very few members who have been recorded, aside from Asgill, is John St. Leger (elder brother of Anthony, founder of the St. Leger Stakes).[309]

In July 1813, Lady Asgill was accosted by a well-dressed man who pretended to be an intimate acquaintance, but being informed that he was unknown to her Ladyship, he at length withdrew; not however till he had snatched a veil from her person, of nearly one hundred guineas value with which he got off, amid the crowd, undiscovered.[310]

Although no longer doing active duty, Asgill was promoted to full general in the army on 4 June 1814.[311]

Charles Ogle's wife, Charlotte Gage, died in 1814, which was a sadness for Sophia's family as well as the Gages. But Charlotte had "not left an inconsolable widower" behind; their marriage had been an unhappy one. Charles was doing exceptionally well in his naval career and was one of the "youngest officers ever made post captain."[312] He was not a lucky man in his private life. Of his three children from his first marriage one daughter died young, and his heir was a wastrel who fled the country in debt, and was estranged from both his father and his wife. Only his daughter, Sophia Ogle, who married an Ogle cousin, seems to have brought him happiness late in life. After her marriage she moved back to the family estate in Northumberland, at Kirkley Hall. The son from his second marriage was classified as a "lunatic," and all three of his wives predeceased him; he was a widower for the last 16 years of his life. One can only hope his career achievements gave some solace. He was the last of the eleven Ogle siblings, dying in 1858.[313] Sir Charles Ogle's life was one of tremendous achievement, and it is obvious that he was a popular, compassionate and highly respected man, who served his country and his fellow men with distinction.

Understanding the Asgills' reduced circumstances, the ever-generous Thomas Graham asked Robert Graham, 12th Laird of Fintry (a Commissioner of the Scottish Board of Excise), to send Asgill a gift of 24 bottles of the best Highland whisky that could be found, sending it to him at 7 York Street, since the original numbering of 6 York Street had changed. It was a parting gift, prior to his departure for Portugal. "It is to be a present from me, therefore nothg. but the freight will come upon him. The best precautions agst *discovery of the contents* of the Box to be taken of course."[314]

As soon as Graham's military career was over (because of eye problems he was plagued with) he immediately joined the Asgills at Brighton, where they were staying at Number 8, Steyne Place. Sophia was so relieved to see him home once more, safe and well. One evening, when Lord Wellington was present, along with Graham, Sir Charles and Lady Asgill were honoured to be invited to the Pavilion by the Prince Regent.[315]

1816 turned out to be a terrible year in terms of losing loved ones. The ever doughty Dame Sarah Asgill died on 6 June, at the grand old age of 86. Once she had been widowed, in 1788, she had moved to 9 Upper Wimpole Street, where she passed away. Her life during 28 years of widowhood had continued much as it had before; she was still admired by politicians and the literati alike. She continued to entertain and keep herself abreast of current affairs. She was a lady who had shown her mettle to the world, back in 1782. Her family mourned her passing deeply. They laid her to rest at St Mary Le Bow, London, on 12 June.

On 22 June 1809 she had written her will:

> This is the last will and testament of me Dame Sarah Asgill of Upper Wimpole Street in the parish of Saint Marylebone in the county of Middlesex Widow first I do hereby give and bequeath unto my dear daughter Amelia Colvile of Upper Wimpole Street aforesaid widow the sum of six thousand pounds and which said sum of six thousand pounds I do hereby direct shall be paid to the said Amelia Colvile as soon as conveniently maybe after my decease and as to all the rest residue and remainder of my Estate and Effects whatsoever and whomsoever after the payment of the said sum of six thousand pounds to the said Amelia Colvile I do hereby give and bequeath the same unto my dear son Sir Charles Asgill Baronet my dear daughter the said Amelia Colvile and my dear daughter Caroline Legge the wife of Richard Legge Esq. in their equal shares and proportions and I do hereby direct that the said third part or share of the residue of my said Estate so given to the said Caroline Legge as aforesaid shall be paid to her or served

to her for the separate use and benefit therefrom and independent of the control or interference of the said husband and not to be subject to his debts or engagements and I do hereby nominate constitute and appoint the said Sir Charles Asgill Baronet and the said Amelia Colvile Executor and Executrix of this my last will and testament hereby revoking all former wills by me at any time heretofore made in witness whereof I now hereunto set my hand and affix my seal this twenty second day of June in the year of our Lord one thousand eight hundred and nine   Signed   Sarah Asgill

This a Codicil to my will I give all my jewels and plate to my dear son Sir Charles Asgill Baronet dated this twenty second day of June one thousand eight hundred and nine[316]

During happier times, the theatre was something the Asgills enjoyed, particularly if it was a Sheridan play. Richard Brinsley Sheridan (30 October 1751 – 7 July 1816) was an Irish satirist, a Whig politician, a playwright, poet, and long-term owner of the London Theatre Royal, Drury Lane. He was also family to Sophia, since Richard's second wife was her first cousin, Esther. "Sheridan was over forty-three and his bride not yet turned twenty, when, on April 27, 1795, he wedded Esther Jane Ogle, the youngest daughter of Newton Ogle, Dean of Winchester ... while she, at first reported to have exclaimed, 'Keep away, you terrible creature,' ended by declaring, so testifies Thomas Grenville, that Sheridan was the 'handsomest and honestest man in England.'" Charles and Sophia were plunged into mourning once more, when Sheridan died, on 7 July, just after they had lost Dame Sarah. They all attended his funeral at Westminster Abbey. "The Asgills were prominent at Sheridan's funeral ... where he is buried at Poets' Corner in Westminster Abbey: personal friends like Erskine and Lynedoch [Thomas Graham], the Dukes of York, Sussex and Argyll followed the coffin. Burgess, Bouverie [Sophia's sister, Arabella's husband], and Asgill followed."[317]

The grief didn't let up during this unhappy summer. Sophia's father, Admiral Sir Chaloner Ogle, died on 27 August 1816. They buried him on 2 September at St. Swithun's, Martyr Worthy, the church at which he had worshiped for so long. It had only been just a few months earlier, on 12 March, that the Prince Regent gave the old officer a patent of a baronetcy at the age of 87. He had not been able to enjoy this honour for very long. Although a somewhat preoccupied father, he had been loved by all his children.

Retirement from the Army

Richard Brinsley Sheridan (30 October 1751 – 7 July 1816). Engraving by Robert Hicks after a painting by Joshua Reynolds dated 1789. Wikipedia Commons.

Although Asgill was no longer a really wealthy man, he did what he could for those with less than him. After the Battle of Waterloo in 1815 a committee was formed, on which he served, to raise funds "for the special Relief and Benefit of the Families of the brave Men killed, and of the wounded Sufferers of the British Army under the command of the illustrious Wellington."[318] His personal contribution was £20, which doesn't sound very much, but it was what he could afford, and was worth so much more in those days.

Admiral Sir Chaloner Ogle, father of Sophia, Lady Asgill. By George Romney, held at the Metropolitan Museum of Art in NYC. Public domain.

Lord Glenbervie dined with the Asgills in April 1818 and recorded the event in his diary.[319] Those present were Barbarina and her daughter Arabella Wilmot, who sat next to the well bred and handsome Captain Stanhope. Thomas Graham, who by this time was Lord Lynedoch, gentlemanly as always towards Sophia, sat next to her that night. Thomas Brand, soon to be Lord Dacre, sat next to Barbarina (who would become his wife the following year). Later Captain Charles Ogle joined them. There were nine of them that night, in their York Street home.

This dinner-party is the last recorded glimpse we have of Sophia. In the last few years her health had worsened, and was the cause of much anxiety to those who loved her. As recently as 1815 she had been taken ill with a fever for a month while at the Lynedoch estate in Scotland. And when visiting the Duke and Duchess of Bedford, at Woburn, Sophia had suffered from "a fit of concussion." By March 1819 they all knew that she was losing her battle for life. Her death, on 30 May, at the age of 48, left a gaping void, and it is said that Asgill was a changed man.[320] She had been 19 when they had married 29 years earlier. Life was never the same without her. She filled a room with her presence and she was deeply missed by all who loved her. On 5 June she was buried in the vault at St. James's Church, Piccadilly, where they had both worshiped whenever they were in London.

Two years later Asgill was attending functions as a single man. His brother-in-law, Charles Ogle, and his sister-in-law, Arabella, Mrs. Edward Bouverie, both accompanied him to the celebration of King George IV's Birthday. Asgill was honoured with a Knight Grand Cross of the Order of the Guelphs, a Hanoverian order, in 1821.[321] Many famous Napoleonic War veterans were included in this order, amongst them the Duke of Wellington, Lord Hill and Viscount Beresford. The Order was founded in 1815 because there were so many to reward after 20 years of war.

In 1821 Charles Asgill sold his York Street home.[322] Money was tight, and the number of staff needed to run the large house was probably not viable for one occupant. On a happier note, the next year it was decided that Thomas Phillips, RA, would paint Asgill's portrait.[323] He wanted his uniform to be in pristine condition for the sittings, since it had deteriorated from lack of use, so he sent it to his tailor to be attended to. He wrote to the tailor from his Pall Mall Freemason's club on Saturday 9 February 1822. He was writing in haste to enclose a draft for the advance payment of repair work he wished to be done to restore his uniform, since it "has become very much tarnished, & will be spoilt unless it is carefully wrapped up." He said he would attend the tailor's premises on "Wednesday at Eleven." Asgill's date did not include the year, but 9 February was a Saturday in 1822. He also failed to mention the name and address of his tailor, which was very likely to have been Gieves at 1 Savile Row, founded in 1771 to meet the needs of the British armed services and is one of the oldest bespoke tailoring companies in the world.[324] He may have been pleased with his portrait, but he never took possession of it himself. This might possibly indicate that he did not like it!

Having sold his London house, it is not clear where Asgill resided in 1822 and 1823. He certainly remained in London, and was active on the social scene. He attended the opera at least five times in March and April 1823.[325] He made an appearance at the King's Levée on 21 April.[326] That evening he was at the Marchioness of Salisbury's Second Assembly, an event that "was attended by nearly the sole of the Fashionable World."[327]

No newspaper indicated that he was accompanied to any of these functions, but he appears to have been involved with Mary Ann Mansel. Her life, and her apparent relationship with Asgill, will be discussed in a later chapter. It is clear that he was at her home at 15 Park Place South in Chelsea on 15 July when he wrote a codicil to his will from there, as he was "now residing here for my health." He wrote another codicil the following day.[328]

Sir Charles Asgill died on 23 July 1823, and was buried with his wife in the vault at St. James's Church, Piccadilly on 1 August. His obituary, carried in newspapers all around the country, carefully and somewhat unusually made no mention whatsoever of his place of death.[329] Upon his death, the Asgill baronetcy became extinct. The wish in his will, written just a month after Sophia's death, was carried out, so that he was laid to rest "in the same vault in St. James's Church, where the corpse of my beloved wife Sophia is deposited."[330]

He bequeathed his portrait to his dear brother-in-law, Charles Ogle, "and at his decease I give and bequeath the same portrait to his son Chaloner Ogle, requesting it may be preserved and retained in his family." Not knowing exactly where the portrait might be located, Admiral Sir Charles Ogle, the sole executor of his will, wrote to Thomas Phillips from Berkeley Square on 23 October 1823 saying that he hoped that the artist will "have the goodness to deliver the picture of the late Sir Charles Asgill to the bearer Mr Goslett" (who was presumably his agent). He also doesn't know if Asgill has yet paid Phillips for the portrait, but if he is owed money he should send his request to "Mr Domville, No. 6 Lincolns Inn", his solicitor.[331]

Retirement from the Army

Sir Charles Ogle's eldest son, Chaloner (the 3rd Baronet), was a wastrel, and fled England owing huge gambling depts. His father disinherited him as a consequence.[332] He died in Brussels on 3 February 1859, less than a year after his father. Charles Ogle's daughter, Sophia (Ogle) Ogle moved back to the Ogle estate at Kirkley Hall in Northumberland after her husband died in 1830. Perhaps she inherited the portrait, and passed it on to one of her six children?

Admiral Sir Charles Ogle, 2nd Baronet (24 May 1775 – 16 June 1858) brother of Sophia Ogle. General Sir Charles Asgill's sole executor and beloved brother-in-law. By Cornelius Durham. Wikimedia Commons.

The Charles Asgill Affair

Retaining the Ogle baronetcy was a tortuous experience for the Ogles, but by hook or by crook it survived a very circuitous route, until 18 June 1940, with the death of the 8th Baronet, Sir Edmund Asgill Ogle (the younger brother of Sir Henry Asgill Ogle, 7th Baronet). Since both bore the name Asgill, perhaps the portrait was handed down, not through the female line in Kirkley, but by the male baronets, as had been Asgill's wish. Perhaps it survived, but at this writing its whereabouts, or its fate, are not known.

Since Asgill had no heirs of his own, it is touching that the Asgill name survived in Sophia's family. Some descendants of his sister, Amelia (Asgill) Colvile, also bore his name, notably:

Augustus Asgill Colvile (1794-1865)

Augustus Henry Asgill Colvile (1841-1926)

Asgill Horatio Colvile (1847-after 1861)

Henry Charles Asgill Colvile (1868-1948)

Charles Asgill Legge (1810-1866) son of his sister Caroline

Robert Asgill Colvile (1916-2008)

Amelia's Colvile family are his only surviving (legitimate) blood relatives, together with the other female line, through Caroline (Asgill) Legge, some of whose descendants live in Canada.

I hope the trend of using the Asgill name will continue, and that he will not be entirely forgotten within the families of his blood descendants. What more can a man ask for, than to be remembered?

CHAPTER EIGHTEEN
Thomas Graham; Lord Lynedoch

Thomas Graham, Lord Lynedoch (19 October 1748 – 18 December 1843) held a deep affection for the Ogle family. It is a story which needs to be told, but Asgill himself was almost certainly not privy to *all* of what follows. Since this very special man and Sophia had a 25 year secret correspondence with one another, it can only be assumed that Asgill knew nothing about it until after Sophia's death, when her letters from Lynedoch were discovered. Unless, of course, his sisters-in-law managed to shield him from the truth.

Lynedoch was described as: "tall, square-shouldered, and erect, his limbs sinewy and remarkably strong. His complexion was dark, with full eyebrows, firm-set lips, and an open, benevolent air. His manners and address were frank, simple, and polished."[333] There are excellent biographies of him, but the focus here is only on his very close relationship to the entire Ogle family.

The Ogle family was large, and consisted of Sophia's father, Admiral Sir Chaloner Ogle, and his brother, The Reverend Newton Ogle, both of whom had married daughters of a former Bishop of Winchester.[334] Chaloner Ogle and Hester Thomas produced eleven very intelligent and interesting children who were, in turn, close with their Newton Ogle cousins. The Reverend Newton Ogle and Susanna Thomas's daughter, Hester, was especially close. Such was the composition of this family when, in 1794, Thomas Graham spent much time with this lively 20-something-year-old family group. Graham, at that time, was a 46 year old widower.[335] He fell in love with them all and they became his closest lifelong friends. On their part, they regarded him as their brother. Quickly, Sophia, the youngest of the girls and the most beautiful of them all, became his favourite – even though she was already married to Charles Asgill. We know all this because of the copious correspondence which passed between them all. Those who knew Sophia had always found her affected, flirtatious and teasing, and it is difficult to reconcile those views with the trust and love he placed in her, especially since his mother would have described her as among "the giddy dissipate multitude."[336]

Placing huge trust in Sophia, Graham confided his thoughts to her, always making her promise to keep his secrets. He expressed views regarding the treatment of his hero, Sir John Moore, and referred to the

King, the Duke of York, Sir Arthur Wellesley (soon to be the Duke of Wellington) and the Cabinet Minister Lord Chatham, and continued:

> yet as walls have ears in London, I have been denounc'd as holding seditious language, & an unheard of attempt was made by government to induce the D. of York to prevent my accompanying the expedition —.[337]

Graham is not at all happy that the suggestion is that he would be capable of mutiny, but points out that his presence already has the Duke of York's blessing.

Sophia Asgill and Thomas Graham had a pact to destroy one another's letters, but in spite of this most of those from Graham to Sophia survived. The only surviving letter written by Sophia, from 1799, no doubt written and kept before the pact was made, appears in Appendix Nine.[338] Biographer Cecil Aspinall-Oglander believes their friendship was never more than platonic, but he cannot have seen *The Diary of Frances Lady Shelley 1787–1817,* in which she describes them behaving at Woburn Abbey with "disgusting familiarity." The letters from Graham, kept by Sophia in violation of their pact, were discovered after her death. While Sophia betrayed this man she loved, certainly as a brother, she provided a service to history, since Graham confided his secrets to her as to no other. But did she betray Asgill? She was the sort of woman who was adored by men, making it inevitable she would have another who wished he could be her lover, if not her husband. But all available evidence suggest that the relationship was platonic – except at Woburn Abbey where it may have gone beyond.

In 1795 Graham was busy with parliamentary and regimental duties,[339] but he found time to spend with Sophia's sister Barbarina and her then-husband Valentine Wilmot. He also enjoyed the company of Richard Brinsley Sheridan, who had also been widowed but was, by then, married to Sophia's Newton Ogle cousin, Hester.[340] He also spent time with Asgill and Sophia in Bath, a favourite watering hole of the aristocracy. By May 1801, Barbarina was writing to Graham to try to persuade him not to go on military service in Egypt, a theatre in the current war, since she feared for his life. The Ogle siblings lost their brother, Major Thomas Ogle (a one-time *aide-de-camp* to General Asgill), at the age of 24, when he was killed at Aboukir Bay that March. In this letter Barbarina told Graham that, were he to stay in England, then he might find "one, who may be something more to you" than a sister.³⁴¹ She ended her letter to Graham with the words "you know how I feel for you."³⁴²

General Sir Thomas Graham, 1st Baron Lynedoch (19 October 1748 – 18 December 1843). Portrait from the frontispiece of his biography by Alexander M. Delavoye, 1880. Public domain.

Who did she have in mind? Herself? Sophia? And what was meant by "more" – a mistress, or a wife? While Sophia and Asgill appear to have had a happy (although childless) marriage, Barbarina's marriage to Valentine was not. They had little in common and were in an unhappy relationship. They separated in 1801 and she lived alone with their daughter, Arabella Jane (Wilmot) Sullivan, for many years.[343] Graham did go to Egypt, despite Barbarina's pleas, and upon returning he asked one of his friends to be, indeed, more than a friend to him. He appears to have proposed marriage to this mystery lady, who, sadly, rejected his offer of more than friendship.[344]

During the Asgill's stay in London, in 1801-2, they were not living at their York Street House because it was being redecorated after their long absence. They stayed at Graham's London home in Stratton Street. When Graham wrote to say he would be passing through and would need to take up residence in Stratton Street too, the big question was, would it be seemly for Sophia to be in the same house with him? Sophia was not a good guide on this matter since "her beauty and charm had always enabled her to do more or less what she liked," nor did she give a damn what the world might have to say on the matter![345] The quandary became so complex that Sophia's sister Arabella Bouverie became involved, making it clear she did not approve of the arrangement and that, of course, it should be the Asgills who turned out and moved elsewhere. She cautioned Graham with *le sage entend à demi mot,* "a word to the wise."[346] She also mentioned to him that the Asgill's house would not be ready for them for at least three months. She told him that she inspected Graham's kitchen the day before and that his cook was in love with him! Thomas Graham seems to have been a man of great charm, honour, and a favourite with women.

As a Bohemian, bluestocking, sculptress, artist, poet, writer, translator, excellent horsewoman and all round good-egg, Barbarina could be relied upon to have the opposite view to her sister, Arabella, writing, "I thought a married woman with her husband might with impunity be under a friend's roof."[347] She told Graham that only more scandal would ensue were he to find lodgings elsewhere and that all of them should stay under his roof. She urged him not to consult their brother, Admiral Sir Charles Ogle, for he had high expectations of his sisters' moral code, and would have agreed with Arabella. She went on to say: "He wants to make 3 Gages of his 3 sisters rather late in the day— besides my good friend tho' we might please him better as 3 Gages we have a better chance of pleasing others as 3 Ogles."[348]

When writing to try to cheer up her gallant friend on 28 February 1802, Barbarina told Graham that she believed it unlikely that Sophia would return to Dublin. A prediction which turned out to be wrong, for Asgill was re-appointed back to Dublin later that year. By July 1803 Graham's regiment, the 90th Regiment of Foot (Perthshire Volunteers), were based in Galway on the west coast of Ireland. Planning to join them there, Graham thought he would first stay with his friends, Charles and Sophia Asgill, who were based in Dublin. From there he would go to Belfast, since the regiment had moved there. He left Dublin about 16 July, which, as it happened, was a week before the rebels planned their Uprising.[349]

Graham's military service was very distinguished, and he earned the high regard of all his senior officers. He would be found wherever he was needed and fought with courage and bravery during the French Revolutionary Wars and then the Napoleonic Wars. He seems to have served in all corners of Europe, the Middle East and even found himself in Scandinavia. It is from Stockholm that he wrote to Sophia, on 18 June 1808, telling her about his exploits, describing the countryside, and mentioning "a draped statue of Pollinia, which I think quite lovely." He ended his letter with the words, "I wish I could help to dress you while I have it fresh in my mind! Are not such thoughts quite out of season and out of character now?"

On his return trip to England, he wrote from onboard the ship *L'Aimable*, on 5 July; speaking candidly to Sophia, he said, "I must recall to your mind how entirely I confide in your not taking any notice whatever of the contents of my letter till at least the subject becomes generally known to the world ... I am extremely anxious not to be quoted."[350] He went on to tell her his thoughts about the exceptionally ill planned expedition in Scandinavia, and that it would all end up being debated and pulled to pieces in Parliament. His commander was Sir John Moore, to whom he acted as *aide-de-camp*, and he emphasised his admiration of the man and the way he handled himself, although he believed Moore would be used as a scapegoat for a disastrous campaign. It is clear he trusted her completely and there is no indication she ever let him down, apart from keeping most of his letters. They had both agreed not to sign their letters, but Graham unfailingly ended his with "God bless you" and "Addio, a.m.," the abbreviation of *amiga mia*.[351]

The Charles Asgill Affair

In 1803, Sir John Moore set up a unique training base for the light infantry at Shorncliffe Barracks, near Folkestone in Kent, in preparation for the Napoleonic Wars. As military historian Gareth Glover maintains: "The scale of light infantry training was really a world first and Shorncliffe could therefore be seen as a World Heritage site."[352]

By 1808 Graham was again in Spain and wrote a very long letter to Sophia on 9 October, complaining that he had not received one from her since he left England. It is full of complaints about politicians, ending with a reminder that Sophia must never breathe a word of its contents, and that he knew she would abide by their pact of destruction. Little did he know! Hardly a day had gone by before he received two letters from her, banishing his gloomy thoughts, and she assured him that she would never forget him. She had written on 12 August and 9 September, and he replied to her on 12 October – a pretty good delivery service for nineteenth century wartime conditions. He said "I shall henceforth refrain from allusions which are useless, but which are apt to force themselves on my paper."

It has often been suggested that, had Sophia not been married, Graham would have taken her as his wife. In his long 12 October letter he told her that he had found a wonderful horse, from the Count Altamira's stable, that he wished he could send to her. He provided her with season tickets to the opera, and told her that if she and Asgill were ever in London, his house at 12 Stratton Street was at their disposal.[353] In a letter of 1 November he told her how badly he felt about being dependent upon the good people of Spain for his sustenance, since they were "suffering under the pressure of war," but he added that he was not enjoying the meals they provided, "an Ebro eel, *stewed in oil*, with garlick!" He said it was pouring torrents of rain, and his mood was shaped by the weather, but he hadn't hanged himself yet, and that his mind was the same as ever. His 5 November letter told her that he was going towards Calahorra and expecting a 10,000-man reinforcement to arrive soon that he hoped this would lead to victory over the French, since he was very concerned for "poor Spain."[354]

He continued to serve under Moore until Moore's death at the battle of Corunna on 16 January 1809, during the Peninsular War. Graham gained his promotion to major general that summer. Throughout all his campaigns he continued to write to Sophia, replying to her letters in detail. She didn't write enough for his liking, and he told her that she was lazy and too occupied with other things, but also of presents he had bought her – a frequent occurrence, with all the Ogle sisters.[355] He took the

Thomas Graham; Lord Lynedoch

opportunity, whenever writing to Sophia, to harangue her with his thoughts and opinions on everyone, and the way war was being waged. For military historians these letters are a treasure trove of information; so frank and honest and so different from official correspondence.

In September 1809, suffering from a fever which had greatly weakened him, Graham was back in London, too unwell to hold his pen when writing to let Sophia know he was home.[356] A year later Graham was writing to her in French, the English translation of which reads: "I do not receive letters. My health is good, but my position is still pitiful. I kiss you." It reads as though his pitiful position was that he loved Sophia, but as a married woman, his dreams of marrying her would be forever unfulfilled. A year later still, having received nothing since hers of 8 December last, he wrote on 20 February 1811 asking her to remember how he felt for her, and that he had done for "so long," adding, "what I have said ... is the *truth*."[357]

After victory in the Peninsular Wars, Thomas Graham became a national hero. His victories and his kindness to all who served under him became well known. "The non-commissioned officers and private soldiers of the Brigade of Guards" wrote to tell him how honoured they were to have served under a man who not only led them to victory, but whose "clemency, noble General, on all occasions, your paternal and indulgent care which made us love you, so it made us follow with enthusiasm your orders, which we well knew would lead us only to glory." For non-commissioned officers of the elite Brigade of Guards to write so to a General, means they knew a great man when they saw one. No wonder Sophia loved him. By volunteering for the army, it was clear he had found his calling. In Parliament, Richard Brinsley Sheridan spoke fulsomely in his honour, saying "never, no never, was there settled a loftier spirit in a braver heart." A Spanish peasant is one of the supporters on his coat of arms; indeed, he called himself "a real Spaniard at heart."[358]

Letters have survived from mothers of Graham's aides which convey appreciation to Graham for the "paternal, considerate manner in which he treated those who served him." Asgill's eldest sister, Amelia Colvile, whose son was on Graham's staff, wrote to the General in emotional terms. Like her mother before her, she was a good letter writer:

> I cannot restrain the effusions of a Mother's feelings by expressing to you my gratitude & warmest thanks for your unremitting care and kindness to my loved Frederick[359]

Amelia's letter is effusive in her praise of Thomas Graham, and her son's love of him. Telling him how much she missed Frederick, she also expressed the consolation she had knowing that he was under his care. Her letter really could have been written by the Dowager (Sarah) Lady Asgill! Perhaps she helped Amelia, since by then Amelia was widowed and the two women were living together at 8, Upper Wimpole Street.[360]

He was always anxious to receive Sophia's letters, which lifted his spirits when they arrived, often more than one at a time. Later in 1811 he told Sophia, "You are the only person to whom I have written (or shall) in this strain," meaning his long diatribes about the tyranny of the French and the "calumnies" heaped on the British for trying to stop their progress. One little episode did cheer him up though, the acquisition of a devoted little white poodle. The dog had lain by the side of his dead French master, General Rousseau, for over 48 hours, refusing all approaches, until Graham sent for him, calling him a prisoner-of-war. Thereafter the dog became his devoted friend, following him everywhere. Graham wanted to send him to Sophia since he knew that this dog must have one master only to whom to show his undying loyalty, but on the battlefield Graham could not provide that security for the poodle. To his cousin, Robert, he wrote: "Since Rover I have never had a dog I car'd so much for".[361]

By June of 1811, Graham had joined Wellington's army in Lisbon. On 7 July 1812 he wrote to tell Sophia of his return to England for eye surgery, which Wellington felt must be done as soon as possible, and gave his blessing, though would miss him as his second-in-command on the battlefield. Graham explained:

> There is a powder prescribed to make me sneeze, wch I take as snuff — most biting to the nostril — the coldest water possible thrown over the head is recommended & generally to keep as much out of the sun beams as may be — [in order to assist his optic nerve, difficult though it is to achieve during active service] ... You wd almost laugh to see the figure I go about in — a blister on my temple and a great green shade like a dowager's bonnet.[362]

Biographer Cecil Aspinall-Oglander writes: "For some unexplained reason this was apparently the last of his letters which Lady Asgill kept. But was certainly not the end of their association."[363] It was while Thomas Graham was on sick leave, in 1812, that he and Sophia met up at Woburn Abbey (Graham was invited to a shoot by the Duke of Bedford), where their behaviour was so frowned upon by Frances, Lady

Thomas Graham; Lord Lynedoch

Shelley. Perhaps thereafter his letters took on a rather different tone, and Sophia knew that she must not keep them.

The story of Sophia and Thomas Graham didn't end there. They were constant companions in London, and elsewhere, after both men in Sophia's life had retired, and we know that she visited his Scottish estate, where the artist Charles Loraine-Smith sketched her Setter dog.[364] Graham's eyesight had become a problem again, otherwise he surely would have been at Wellington's side at Waterloo. Sophia "had played her hand so well that his devotion to her was as complete, and as undemanding, as ever. Even Sir Charles could accept the situation without a qualm, and the two men were on the best of terms. Sophia was evidently a clever as well as an attractive woman."[365]

The Asgills were in his company when he went on grand tours of Paris and beyond, for a three month's tour, at his expense. Having not selected well the servants who accompanied them, Asgill, "who requires a deal of attendance, is out of all temper."[366] A very ungracious attitude, since the servants were not being paid for by him. Graham arranged for them to visit King Louis XVIII, and another time a visit to the theatre, "where Lady Asgill had been given the Duke of Wellington's box."[367] The Duke had lent them his box at the Comédie-Française to see the great Talma in Corneille's *La Mort de Pompée*. One theatre goer described Sophia as "the strangest and least describable of all demireps," and Arabella as "her elegant and amiable sister Mrs. Bouverie"; he called Asgill "a true Sir Charles Easy of a husband," and Graham "the heroic, chivalrous, most gentlemanlike, and much enduring lover of Lady Asgill."[368] Was his language flowery, or did he actually believe, like Lady Shelley, that they were lovers?

In 1817 they all did another tour, this time to Belgium and France.

Graham was awarded a Peerage in 1814, henceforth to be known as Lord Lynedoch.[369] The Ogle sisters, Sophia, Barbarina and Arabella, got together to design his coat of arms, with Barbarina, the well-known horse sculptress, drawing the rearing horse as one of his supporters.[370] He became a full general in 1821, at the age of 73.[371]

Arabella wrote to tell Lynedoch that he certainly should accept the pension which went with the barony (which he had intended to decline) and told him about Sophia's health: "I have not heard S. complain of her head anything the least resembling what she did. Her complexion is a good deal heated, I believe she thinks herself the better for it, it is not becoming, and therefore I will not call her in good looks." To which Barbarina added a characteristic note: "If you will come home all will be well, but how will you contrive to amuse yourself without fighting is the point."[372] Sophia's symptoms, in Ireland, had been chest-related, but here her sisters described headaches and facial flushing. The Spring of 1819 saw Sophia's death draw near. "For the last five or six days I have not stirred out of Sir Charles Asgill's house," Lynedoch wrote on 12 May, "and we have all been in the greatest alarm about Lady Asgill, whose health for some months has been very indifferent, but a sudden and violent attack of spasm has put her life in imminent danger and has left her in so low and reduced a state as to render the result very precarious."

Only days after her funeral, it was discovered that she had not, as promised, destroyed Lynedoch's letters. "A large packet of them had been found amongst her treasures, and they were now returned to him, apparently by one of her sisters."[373] It is comforting to know that Lynedoch then had a record of what he had said to her, records he obviously considered worthy of safe-keeping for posterity. One can hope he didn't feel betrayed, and came to the conclusion that she just couldn't part with them, out of love for him. As the Scottish Judge, Henry Cockburn, wrote of Lynedoch, in 1837:

> Nor has it been only in the affairs of war that his manly chivalrous spirit has made him be admired and loved.[374]

When family and colleagues started to see a decline in their friend's health, many worried about him, including Queen Victoria; regularly, every morning, the Queen enquired how Lynedoch was faring, and wrote to tell her uncle, the King of the Belgians, that she feared the worst (as was reported in *The Times* on 18 December).[375] That same day he died at his Stratton Street home in London, in the 96th year of his age, after a very short illness: indeed, he rose and dressed himself on the day of his death.[376]

Lynedoch's death, nearly a quarter of a century after Sophia's, was a terrible blow to the remaining Ogle siblings. Barbarina mourned his loss greatly, since they had been lifelong friends and frequent horse-riding companions, well into their old age.

CHAPTER NINETEEN

The Swindler Asgill

Six weeks after Charles Asgill died, *The London Gazette* carried an announcement that was simple and to the point: "ALL persons having claims upon the estate of the late General Sir Charles Asgill, Bart, are requested to send an account thereof to Messrs. Graham, Kinderley, and Domville, Solicitors, Lincoln's-Inn, London."[377] Admiral Sir Charles Ogle, Sophia's brother, was the sole executor of Asgill's estate. In addition, Asgill had not lived long enough, or possibly not been well enough, to administer Sophia's affairs, and an estate of £3,000. This too was dealt with by Charles Ogle, as executor.[378] The will, written soon after Sophia died in 1819, concluded:

> And I hereby lastly give and bequeath unto the said Sir Charles Ogle his executors, advisors and assigns all my plate, books and jewels and, revoking all former wills by me at any time made, do declare this to be my last will and testament. And I appoint the said Sir Charles Ogle sole executor thereof, in witness whereof I the said Sir Charles Asgill have to this, my last will and testament contained in six sheets of paper at my hand and seal, that is to say my hand to the five first sheets and my hand and seal to this sixth and last sheet this thirty first day of July in the year of our Lord one thousand eight hundred and nineteen. Charles Asgill. Signed, sealed, published and sworn by the said Sir Charles Asgill, Baronet, the Testator, as and for his last will and testament in the presence of us, who in his presence at his request, and in the presence of each other have hereunto subscribed our names as witnesses thereto Wm Domville - Cadogan Morgan – Achibald Thomson – Lincoln's Inn.[379]

The General had scribbled codicils out on his own just days before his death, with no solicitor present, and only witnessed by his physician, William George Maton M.D. Much debate took place to ascertain that these codicils were genuine and the writing recognised; in with his will were affidavits to this effect, ending:

> And the said William Domville for himself further made oath that the said deceased departed this life on or about the 23rd day of July last and that on the same day he went to the Banking House of Messrs Basset, Farquhar and Co, the said deceased's

Bankers, and there found the Will and first Codicil of the said deceased sealed up in an envelope and on the same being opened he observed the words "and I give to the" and "my portrait painted by Phillips" written on razure in the said twenty fifth and twenty sixth lines of the said, and an razure at the beginning of the twenty seventh line of the said sheet, so that the said line contained only the words "and at my demise I give and bequeath" and that the said Will and first Codicil are in all respects in this same plight and condition as when so found

Wm Domville – Geo Streatfield – on the Fifth day of August 1823 the said William Domville and George Streatfield were duly sworn to the truth of this affidavit before me J Addams Surr. Pros. Richard Addams Not. Pub.[380]

Since Asgill's brother-in-law, Charles Ogle, was his sole executor, he had to deal with finalising the General's estate and that of his sister, Sophia's.

One of the codicils, written on 15 July 1823, eight days before his death, concerned money that he had apparently spent for his regiment for which he was owed reimbursement: "The sum of eight hundred pounds which are at my agent's hands, Messrs. Greenwood & Co., Grays Court, agents, or will be deposited this month at my off reckoning, for the cloathing of the 11th Foot, due to me."[381] Was this the money he had spent raising the regiment's second battalion a decade and a half before? Was this the reason he had to sell his York Street house, on account of a shortage of funds?

There was other money from the past that was still in limbo. When Asgill was being held at the tavern in 1782, Major James Gordon secured £500 from the British paymaster general to fund food and other needs for Asgill's comfort. The money had never been ratified, so the paymaster would be responsible for recovering it. After the war the British government was faced with an enormous number of financial claims and attempted to recover every possible shilling. This money was, eventually, demanded to be paid back by Asgill personally. He never paid it, arguing that it had been a necessary cost to save his life. Many years later Samuel Graham, one of the officers who had drawn lots with Asgill in 1782, received a letter from the former paymaster general:[382]

You will be surprised to hear that Asgill refused to pledge himself for the repayment of the monies I advanced him (in case government disallowed the charge), and that a very unpleasant correspondence passed between us on the subject, which is still in my possession.

He closed by saying that the money was, finally, written off by the government.

Asgill's eldest sister, Amelia, outlived him, but by less than two years. Her marriage to Sir Robert Colvile had been short-lived, just thirteen years; he died in 1799. She and the Dowager Lady Asgill lived together after that, companions in their widowhood. Amelia died on 12 July 1825.

Barbarina's estranged husband died in 1819. Before the year was out, on 4 December, she married Thomas Brand, the new Lord Dacre, the poet, playwright, and translator.[383] This time she was in a happy marriage. In addition to her writing, she rode, was proficient in both French and Italian, and maintained an extensive correspondence with a circle of other literary women.[384] She lived until 17 May 1854.

Sophia's eldest living sister, Arabella Bouverie, was widowed in 1825. Three years later she followed her sister Barbarina's lead in marrying a man much younger than herself, the Hon. Robert Talbot. She lived until 29 October 1855.[385]

Asgill's youngest sister Caroline's later life was especially sad. She had married Major General Richard Legge (Royal Irish Artillery) on 5 April 1800, and they both seemed happy together, but Richard was an inveterate gambler and inebriate, which must have caused problems in their marriage. He died on 12 June 1834 at their home of Ninnage Lodge in Westbury-on-Severn, Gloucestershire. Five years later, on 23 August 1839, Caroline lost her youngest son, William Legge; he put a gun to his head, on account of a woman of whom his mother, Caroline, disapproved.[386] He was just 20 years old. She lost her mind and was admitted to the Gloucester First County Asylum on 2 September 1839, as a 1st class resident, number 977 (although she had been "without lucid intervals, from June 1837").[387] At the end of February, 1840 the following articles appeared in the press:

> Commission of Lunacy.— Thursday last, a commission was opened at the King's Head Hotel, this city, before Dr. Maddy, Mr. Mortimer, and Mr. Addison, to inquire into the state of mind of Mrs. Caroline Augusta Legge, of Westbury-upon-Severn, widow

of Genl. Legge, and the mother of the young man who some time since committed suicide at Westbury. The lady is at present an inmate of the Gloucester Lunatic Asylum. The facts brought forward in evidence with the view to establish her insanity are such a nature as to make it totally impossible even to allude to them in a public paper.[388]

What on earth could be the situation that it had been impossible to even allude to them?

Commission of Lunacy. — The inquiry into the state of mind of Mrs. Legge, widow of General Legge, of Chaxhill, in this county, was continued Saturday and Monday last, upon which days the alleged lunatic was personally examined by the Commissioners and the Jury. Further evidence was also adduced, which, with that previously given, is of a nature unfit for publication. The Jury returned a verdict to the effect that Mrs. Legge had been of unsound mind, without lucid intervals, from June 1837.[389]

Legge's widow, Caroline Asgill Legge, the daughter of Sir Charles Asgill, first baronet, and the youngest sister of Sir Charles Asgill, second baronet, died at Ninnage Lodge in 1845. The estate was inherited by their son, also Charles Legge, who in turn settled Ninnage Lodge on his daughter, Caroline Legge, on her marriage to Thomas Goold of Newnham in 1864.[390]

Immediately after Charles Asgill's death, the Asgill name was again all over the newspapers. Besides the death notice, various tributes, notices of who would succeed him as colonel of the 11th Regiment of Foot, and the announcement that the position of Equerry to the Duke of York would now remain vacant, a mysterious new name appeared.

Around September 1823, according to newspaper accounts with the headline "Extraordinary Adventurer," a clergyman in Shaftesbury, Dorsetshire, advertised in a newspaper for "pupils of respectable connections" to reside with him. He received a letter "purporting to be signed by Sir Charles Asgill" proposing to place his nephew with the clergyman; the letter claimed that this nephew "had been to college and subsequently in the army." The clergyman determined that Sir Charles Asgill was "a person with whom he might, with propriety, form a connection" and agreed to take the student for a fee of £300 annually.

The Swindler Asgill

The pupil arrived – a shewy, talkative, impudent young man, about five or six and twenty. For the first two months he was seldom seen about the town. Nothing could have been more politic. By degrees the stranger was more indulgent to the curiosity of those who felt any desire to make a nearer approach to the Baronet's nephew. He dined out frequently, charmed the ladies with small talk, pleased the gentlemen with his easy vivacity and knowledge of the world, filled the learned with admiration by his readiness in quoting Horace and capping verses, and by talking familiarly of Eton and Oxford, and became in short the general favourite. His liberality too was remarkable. Champaigne and claret flowed without stint for the enjoyment of his friends; and even the passing guards and coachmen were made happy by his bounty. His dress was of a splendour corresponding with his general habits of magnificence. People not only opened their houses, but their purses. Mr. Asgill not only got what he wanted, but did some persons the honour to borrow money from them upon his acceptance.[391]

Then he disappeared. This raised a hue and cry in Shaftesbury, and various people came forward who had been lured into the trust of this supposed nephew – a tailor, a silversmith, an innkeeper. The man's deceit was thorough: he had regularly sent letters to Sir Charles Asgill at a hotel in London, and provided others with an address for him at a country house; now that he had disappeared, the creditors found that Sir Charles had never stayed at that hotel, and that there was no such country house.[392]

Reports of this alleged Asgill nephew began appearing in newspapers in late August and early September, coincidentally soon after Sir Charles's death. Quickly more people came forward with more stories of being bilked for goods and services. By the end of the month, a more comprehensive report had been compiled that was published in a number of newspapers, including *The Observer* (London) of 8 September 1823 under the headline, "Imposter cashes in on Asgill name." Others read:

THE SWINDLER, ASGILL

We have hitherto abstained from inserting any account of the infamous, though highly accomplished swindler who has been for some time resident within 20 miles of this city, until we could collect our information from the most correct authorities. Last January a letter, bearing date from Aberystwith, and purporting to be written by Sir Chas. Asgill, was addressed to a clergyman in this neighbourhood, who receives pupils. It stated that the Baronett had a nephew, a "young man of superior attainments, gentlemanly manners, and amiable disposition," who had received a highly finished education, but having been in the army for some time, had in a great measure forgotten his classical knowledge. Having a valuable living in Norfolk, Sir Charles expressed his intention of presenting it to his nephew; and wishing to procure his ordination without the formality of a University education, it became necessary that he should be placed in some such situation as the one to which application was then made. An answer was returned, in which some doubts were expressed as to the probability of a young man, accustomed to the gaieties attending a military life, conforming to the retired habits of a clergyman's family. This objection, however, was entirely removed by the succeeding letter, in which the clergyman was assured by Sir Charles, that although his nephew was of an age which required his being considered more in the light of a companion than a pupil, yet his disposition was so amiable, that he would be found in every respect willing to conform to the regulations and plans adopted by the family. In the same letter persons of the highest respectability, in Somersetshire, known to the clergyman by reputation, though not personally, were mentioned as having been consulted as to the situation of his residence, by which means no suspicion could be entertained of a fraud; on the contrary, the respectability of the connection seemed unquestionable. Some farther preliminaries were settled through the medium of the post, during which correspondence the worthy uncle made known his intention of personally introducing his nephew to his new place of residence. The interview, however, was cleverly waived by the convenient illness of his friend, Lord — of —, Worcestershire; from whose house the excuse was supposed to be sent; and it is a remarkable fact, that the nobleman named was at that time seriously ill; and shortly after died. To trace, step by step, this extraordinary adventure, would be useless; suffice it to say, the dashing nephew arrived – talked largely of his acquaintance – claimed for his cousin a Baronet, whose name he has lately honoured by assuming – was quite happy to find his connections so well known; and even informed the family of a lucky change of name and fortune which had lately befallen one of their relatives, which at that time to them was unknown. The farce was kept up by a list of books being ordered, and the embryo of a Divine duly attending in the study every morning. For some weeks every thing answered the flattering description: in all things he was obliging, consistent, and considerate; but suddenly he was seized with an irresistible inclination to visit London for a few days. Preparations were accordingly made, and he himself politely offered to be the bearer of any communication to Sir Charles, to whom a flattering and satisfactory letter was dispatched.

In a few days a distressing account arrived from his uncle, in consequence of the military disposition still evinced by his nephew, and which he hoped would have been eradicated by the change in his pursuits and companions. In a postscript, the enraged uncle did not forget to request the clergyman by all means to suit his own convenience in drawing upon him for his payments, which however was declined, until the expiration of the half year, before which time the hopeful pupil had decamped, under a pretended order from the War Office. On his return from London, a strange metamorphose had taken place, an astonishing disposition for expense appearing in every department of his finances – two horses, a stanhope, and manservant were kept.

An application from a gun-smith and jeweller respecting the fortune and family of this gentlemen, who had given them large orders, caused no small uneasiness to the clergyman; who, fearing that his pupil's generous disposition was leading him into expenses, which his income, however respectable, would by no means warrant, returned such answers to these applications as induced these trades-people to decline the honour intended them. It would have been a fortunate circumstance if those who have suffered by his artifice, had adopted the same prudent principle of not supplying goods without making proper enquiry. Had they done so, in all probability this extensive fraud would have been detected in time to have stopped the conduct of the swindler, and have brought him to condign punishment. These circumstances led to many serious conversations with the apparently thoughtless young man, and finally to a confidential correspondence with Sir Charles, who requested that no particular might be concealed from his knowledge. He expressed his fears that all had not been right from the tenor of his nephew's letters of late; at the same time observing, that his nephew's income, the pay he received, and an occasional present from himself, were abundantly sufficient to defray any expenses he seemed already to have incurred.

This communication respecting his expenses, had the effect of placing the swindler more on his guard, as from that time he received either in private or at some other house, the goods which he was constantly in the habit of obtaining; nothing but a few of baubles being exhibited to the family.

Fresh causes for disapprobation now began to appear in his conduct, which called forth severe remonstrances on the one hand and firm assurances of amendment on the other, which were however forgotten as soon as made.

This produced a determination in the clergyman of having an interview with the uncle, but at the time appointed an unlucky fit of the gout rendered it impossible for Sir Charles to travel. Three painful efforts were made by the invalid to write on the subject, which was at last accomplished in a letter almost unintelligible. His gratitude for the kind interest evinced toward his nephew was expressed by most cordial invitation to the family to pass some weeks during the summer in Wales. This invitation was extended by the nephew to many of his friends, and he appeared to enjoy the idea of having so large a party to partake of his hospitality at the manor house of Aberystwith.

But a farce cannot hold out forever; and this having existed for almost six months, was necessarily now drawing to a close. One more letter, and that of a most serious import, was written concerning him. It urged his removal to some other situation in consequence of his conduct being such as to defeat the purpose of his benevolent relative. Very shortly after, as might be expected, he took a gracious departure, under the pretence already mentioned.

So well was the fraud carried on, that not the slightest suspicion seems to have arisen in the mind of any individual as to the identity of his person. Even after the death of Sir Charles Asgill, this audacious villain held a correspondence with a most respectable gentleman in the neighbourhood, where he had been playing off so profitable a game. An application being made to the executors of Sir Chas. Asgill for the sum due, as was supposed, from the estate, a polite answer was received, expressing some astonishment as to the person who had been received as a nephew to the Baronet. A hasty journey to London was the effect of this answer; when, to the no small chagrin of the clergyman, he was assured, by those who were best able to judge, that the letters he held so confidently in his hand had no similarity to the writing of Sir Chas. Asgill. The reputed nephew was even unknown to that family, and the most daring fraud had been perpetrated. A coincidence in events appears in every instance to have favoured this singular plot. Sir Chas. Asgill's absence from the opera, according to the public prints, was always well timed with the accounts of his attacks from the gout. His last severe illness and subsequent death were still further corroborations of what had passed and served to lull all parties into the utmost security. And here an opportunity occurs for mentioning the manner in which the correspondence so often alluded to with the supposed Sir Charles was carried on. By an adroit manoeuvre he obtained possession of all the letters, and answered them himself. The first two sent to Aberystwith, he was on the spot to answer, as he has been traced into Wales about that time. Those written after his arrival in this neighbourhood, were directed to a respectable hotel in Bond-street, and these he secured by the following method. A few days previous to his commencing his operations in this disgraceful career, he took up his abode at this hotel, where he lived in style. At his departure, as it were *en passant*, he informed the master of the hotel that as his uncle Sir Charles intended using his house in the spring, it might happen that an occasional letter would be directed to him there before his arrival. At present he was in the country on a visit to an old acquaintance whom he had not seen since his return from India. But his movements were so uncertain that it would be the better plan always to send these letters under cover to him (the nephew), giving as his direction the residence of the clergyman where he afterwards took up his abode. The hotel-keeper, unsuspicious of his new customer, attended to these instructions, forwarded the letters, which were answered as best suited the purpose, and through some accomplice were put in the London post. As they were not letters which required an immediate answer, the lapse of 4 or 5 days was a circumstance unnoticed by the clergyman, particularly when he considered the ill state of Sir Charles's health. The three specimens which have been traced to be his writing, are so totally different in their character, that no one would imagine them to be the writing of the same person. Unabating exertions have been made for tracing the swindler and his connections. Unfortunately they have failed in discovering his present retreat, and it is much feared he has eluded the grasp of justice. By birth he is found to be of high respectability, but his extravagance and ill conduct have induced his family for some time to disown him. His abilities are of a superior cast. At an early age he distinguished himself as the head boy of a public school, from which he was removed at the age of sixteen to one of our Universities. We are happy to find that the extent of his frauds are much below what is generally reported, though we are sorry to add that many poor and industrious trades people will suffer severely.[393]

Who was this mysterious swindler, using the name of Charles Asgill, claiming to be the nephew of the recently-deceased Baronet? None of the keywords, like Norfolk, Aberystwith, or India, (and one account had Asgill living in Regents Park, equally inaccurately) have ever been associated with Charles Asgill, nor any evidence of gout been found. One newspaper mentioned the name "Armitage," but proof would be needed to establish that this very clever, scheming man was him.

PART II

The Charles Asgill Affair

CHAPTER TWENTY

William Charles Asgill

Precedents for scandal, mistresses, gambling and debt were set by several members of the family of King George III, much to his dismay. The Duke of Clarence was no exception and his mistress, Mrs Dora Jordan, "bore the Duke ten children, all of whom took the name Fitzclarence."[394] Clarence succeeded his brother, George IV, to become King William IV on 26 June 1830, until his death in 1837, when his niece, Victoria, became Queen. She was the daughter of the Duke of Kent (4th son of George III). There were many illegitimate grandchildren from other sons. Jordan's statue is on display at Buckingham Palace. Another scandal, of many, concerned Asgill's ultimate boss, the Duke of York and Albany, and his mistress, Mrs. Mary Ann Clarke. Not only was there the scandal of cash, in exchange for promotions in the army, instigated by Clarke, but the Duke "born in 1763 filled his days with wine, women and gambling, plunging headlong into debt"[395]. Had he not died in 1827, York would have succeeded to the throne after George IV. These scandals formed the backdrop to Asgill's life, so perhaps there is no surprise to know that he followed suit, certainly with regard to (unsuccessful) gambling.

Having told Charles Asgill's story, now I will tell mine. They meld together, since for 20 years I've done little else but live with Charles Asgill as the central focus of my life.

In late September 1998 my father, Colonel Philip Newton, died. Following his death my step-mother offered me the contents of his study, in which, when unpacked from the black-sacks in which they were handed over, I discovered his genealogy files. Hundreds of pages of work done on his American mother's ancestry, prepared jointly with his eldest brother, Professor Roy Newton, spilled out. It was all fascinating. Much had been achieved during his appointment as UK Representative at the United States Command and General Staff College (Fort Leavenworth, Kansas). While there, all his free time was spent on work which eventually enabled my grandmother to become a Daughter of the American Revolution – a prestigious organisation, which takes ancestry very seriously.

My grandmother was born Thelma Celeste Hammond, at 33 Cherry Street, Indianapolis, Indiana, USA on 26 July 1893. She married an Englishman, Henry James Hall Newton, on 6 September 1911. Living in England, she was thrilled to learn from my father that her heritage went back to several notable early American settlers. One was the Hicks family, who owned a very large farm in Brooklyn. Hicks Street there is named after her ancestors. When the pressure to build houses overtook the need to farm, they sold up and became ferry operators between Brooklyn and Manhattan.

His Majesty, King George III of Great Bretain and Ireland (4 June 1738 – 29 January 1820). By Johan Zoffany, 1771. Wikimedia Commons.

William Charles Asgill

One fascinating story turned up. One of her – and my – American Hicks ancestors, Captain Robert Hicks, was an ardent revolutionary. To protect his assets in those uncertain times, he buried a pot of gold and silver coins on one of his properties. He died before he had told anyone where he had buried it, so his family were unable to retrieve it.[396] So, you see, I have ardent Patriots as part of my DNA.

Back in 2002 I had only recently acquired my first computer, so I downloaded software which enabled me to log my father's work into a programme which laid it out well. From unpacking the black-sacks, to surveying a broader screen-view of it all, took some time, but it was clear that it was very heavily weighted on my paternal heritage. I had to do something to balance the scales as it were. I discovered an Ancestry.com online course which I hoped would help me learn how to do my own search.

I really knew nothing of my mother's mother's ancestry, but knew her maiden name was Edmed. Janet Reakes, a genealogist in Queensland, had taught us the value of One Name Studies. Lo and behold, there was an Edmed One Name Study! The creator, Jim Edmed, was generous and sent me all the information, going back to the mid-eighteenth century. What a boon – I didn't have to do the donkey work. Of course, this only led me to my mother's mother's father, and beyond. Now, what was the story of my mother's mother's mother? My mother, Joyce (May) Newton, was still alive, so on my next visit I asked her. She knew nothing at all. So it was lucky for me that Jim Edmed had told me how to contact a genealogist (and a third cousin) named Kathleen Luckcuck, telling me she had an incredible story to tell, in which General Sir Charles Asgill featured prominently.

She did indeed. We arranged to meet up, after Kathleen had sent me all the basic information, and we did so at Asgill House on the weekend of Queen Elizabeth II's Golden Jubilee, 4 June 2002. Driving from the south east of London, through to the west, was simply awful; the traffic was bumper to bumper. My daughter, Nina, travelled down from Norwich, where she was at university, and her journey into London and beyond was appalling too. In spite of having logged all Kathleen's work into my database, making my mother's ancestry look a little more promising, she and I were in for a bombshell when handed letters by the owner of the house.

The Charles Asgill Affair

Fred Hauptfuhrer, the owner of Asgill House, brought out a trove of letters written by members of the Asgill family, giving details of their heritage down from William Charles Asgill, the son of Charles Asgill. I clearly remember the look Kathleen and I gave one another on hearing that news. Astonishment. Sir Charles Asgill died without issue, so far as either of us knew. And yet here was a family history beginning with his son. After spending time with Fred, who kindly showed us some of the house, we went our separate ways. Kathleen later went to the National Archives in Kew to find out who these Asgills were. She found one descendant of William Charles Asgill, but I think that was all. Looking at those letters again revealed that a descendant of the said William had emigrated to New Zealand.

The search was on. I put a "wanted" article in a New Zealand newspaper, in Napier, and some information was relayed back to me from that. One kind lady who had known the Asgills sent me an entire photograph album. Eventually it became clear that the modern-day Asgills were in Scotland, and the name of a nursing home in Edinburgh was given to me. I was nervous. Would this be the breakthrough I needed? My former boss, Peter Drummond-Murray, lived in Edinburgh and he kindly agreed to deliver my letter personally to the address I now had. He did so very promptly, but was greeted with the news that the lady who had been at the nursing home had died recently. Eventually my letter was delivered to her daughter, Jennie Young. When we finally made contact, it was a "Livingstone, I presume" moment! I felt I had been on a long and arduous exploration, through deepest, darkest, Africa, and couldn't believe it when I heard her voice on the telephone.

Jennie was kind and generous with all the information and paperwork she shared with me. I was given the groundwork of an incredible story of disinheritance and marrying beneath one's station. Her ancestor, William Charles Asgill, had been the son of Charles Asgill and had been expected to become the 3rd Baronet, and Asgill House would be theirs today had not General Sir Charles Asgill disinherited William. I was even given a copy of William's obituary, which read: the "second son of the late Charles Asgill"[397] – stating the latter was "of Regents Park." Asgill never had an address in Regents Park. By now I had tracked down Jennie's cousin, Andrew Morrison, whose mother had been the twin of Jennie's father, whose wife had died in the nursing home in Edinburgh. I was really into the core of this story now, since Andrew had done an enormous amount of work on the ancient Asgills, and had found the heritage of Sir

William Charles Asgill

Charles. Once more, the genealogy work was handed to me on a plate and it was a few years before I went to Kew's archives myself.

William Charles Asgill's story had been passed down by Jennie and Andrew's ancestors, the sons and sons of sons, down to the present-day generation. It seems William was an educated man who eventually became a schoolteacher, owning his own establishment in Beddington, Surrey, where he taught maths and Latin, bringing up his children with his wife, Mary Leutchford. Apparently Mary was the problem, since she was a miller's daughter, from Sevenoaks in Kent. Sir Charles Asgill deemed her a totally unsuitable bride for his son; hence his disinheritance. It is not known how William met Mary, but they fell in love and married at St. George's Church, Hanover Square, London, in the autumn of 1823, just a couple of months after Sir Charles had died, and could therefore do nothing to prevent it happening. It was a society church then, where notable people married and had their babies baptised, although I did hear that it had a slightly "Gretna Green" side to its history too.

So I knew a little about William Asgill now. *I was also told that he had wanted to become a clergyman,* but had disagreed with one of the tenets of the church, so he changed his plan and became a schoolteacher. But there was now an elephant in the room: if William was "the second son," who on earth was the firstborn son of Sir Charles Asgill? We found a "Henry Asgill," buried in Winchester, and lit upon him as a possibility for a while since his age seemed to fit in with the story.

My attention had been diverted by hearing William's sad story, and I did little work on my own and my mother's heritage at that time. By now I had embarked on the first of several house moves in Spain and Gibraltar, which had also been disruptive to orderly research. I felt so sorry for William though. Andrew Morrison and I became a team, and we both worked tirelessly to try to establish the bedrock of William's story. It seemed to elude us at every turn. We thought "The General" (as we called him to distinguish him from his father and uncle, both also named Charles Asgill) would have sent him to Westminster School, where he and the Lord Mayor had been educated. We drew a blank there. We then searched for him as a pupil at all the other well-known public schools of his time. More blanks. *We tried to find evidence of him undertaking religious studies.* Blanks again. We searched for William high and low, but could never find him beyond the family's handed down story, which told us he had been born circa 1800, in Ewell, Surrey. But he appeared in no church baptisms whatsoever in the region around that date.

The Charles Asgill Affair

A while later, after yet another house move (or two) on my part, Andrew found William in Debtors Jail. With that discovery we began to think that William had been an illegitimate son of The General since, we believed, his aunts (The General's sisters Amelia and Caroline) would not have allowed that to happen to him for the sake of the Asgill reputation. We then discovered that even if he had been disinherited, only an act of parliament could deprive him of the baronetcy. William's story was looking to be on very shaky ground, indeed no bedrock at all. We were perplexed, disappointed, and very sad. We resorted to the DNA route (the story of which I will come to), but there was no match. By this time I had spent several years of my life not only believing William's story and trying to find him, but becoming very fond of him. My mind was thrown into disarray at the lack of a DNA match. It affected me badly and interrupted my concentration on the overall genealogy work which needed to be done.

Some considerable time later I found the newspaper reports of "The Swindler Asgill." Everything reported about the swindler fitted William's story. Well educated, personable, and especially the religious training, which the swindler was pursuing – too many coincidences to be coincidence. But now he was a "nephew" of Sir Charles Asgill, who had apparently given him *carte blanche* to swan around the country incurring huge bills for purchases made, on the strength that the vendors only had to send their invoices to The General for settlement. All this happened in 1823, in the six month period leading up to The General's death, as was reported:

> His last severe illness and subsequent death were still further corroborations of what had passed, and served to lull all parties into the utmost security.[398]

I have never found evidence of what The General's reaction to receiving these substantial invoices may have been. It was reported that his executors denied all knowledge of a "nephew" by the name of Asgill, because there simply wasn't one. The General had no brothers.

As I write, we have reverted to the DNA route to find out who William Charles Asgill, *a.k.a The Swindler Asgill*, really was. He was born by a totally different surname, even if the forenames were the same, so it was no wonder we never found him anywhere, other than in the Debtors' Jail. To our utter amazement, after I wrote to Professor Turi King at the University of Leicester (famed as a consequence of her discovery of King Richard III in a car park there) she wrote back to me showing some interest

William Charles Asgill

in following this DNA trail. As a consequence, she said she would need another DNA test done for Robert Asgill (a Y-DNA test) since Robert is a direct male-line descendant of William Charles Asgill. The test has been done, but the whole process of discovery will take some considerable time. If not with the help of Turi King, then concentrating on the genealogy route.

CHAPTER TWENTY-ONE
Charles Childs

For quite some time, Andrew Morrison and I believed Charles Childs and William Charles Asgill were half-siblings (both being illegitimate sons of The General, or so we thought, for several years). I still regard Andrew as my cousin, even though DNA has proved we have no blood link to one another. Some areas of what follows have been verified, but some have been impassable brick walls. I cannot offer proof, but there is a huge amount of circumstantial evidence, nonetheless.

Kathleen Luckcuck's letter to me, back in early 2002, had, indeed, been an incredible story. The fact that we had stumbled, inadvertently, on the story of William Charles Asgill, was absolutely not expected by either of us, hence the "knowing look" we gave one another at Asgill House. Kathleen had written to me about hers and my descent from General Sir Charles Asgill through his illegitimate son, Charles Childs. Like Andrew Morrison, with his hand-me-down family story, all I could do was to believe the findings handed to me by Kathleen. She had been delving into it all for several decades.

She told me that Charles Asgill, The General, had a mistress by the name of Mary Ann (Goodchild) Mansel. The Mansel surname was an alias she had chosen for herself, which she used to cover her tracks as the mistress of General Robert Manners, a grandson of the 2nd Duke of Rutland, and an Equerry to King George III. Mary Ann was in a long-term relationship with Manners and they had six children together, all baptised at St. George's, Hanover Square (where The Swindler married Mary Leutchford). Their sons, William Henry and Henry Edward, died young. At the time of Mary Ann's death, only three of their six children were still alive, Robert, Lucy and Herbert, a son George also having predeceased her.

I tracked down descendants of Robert Mansel in New Zealand, and Herbert Mansel's descendants were not far from where I was living at the time. Lots of communication between both families ensued. Herbert's family lent me many letters written to Herbert in the nineteenth century, and allowed me to transcribe them all prior to returning them. The letters were very hard to transcribe. They had been bundled very tightly, while folded, and the text had vanished from the folded lines.

A letter from Lucy Mansel (with a P.S. from Mary Ann Mansel) to Herbert Mansel (Charles Childs' closest sibling). Author's collection.

The post script from Mary Ann Mansel which continues on the same letter from Lucy Mansel, dated 29 May 1833, reads:

> My Dear Herbert, How much I wish you were here, that I might nurse you. Surely they will allow you leave of absence to perfect your recovery. This place is now looking beautiful. I hear all is

more quiet and you might get away I should hope. Lucy has said how soon I am to lose her. God Bless you. Pray let us hear from you very soon. Your affectionate Mother, M A Mansel

These Mansel letters were mainly from Herbert's siblings, telling him of family tittle-tattle and of life in the army for the two eldest brothers. However, two or three stood out. One was from General Robert Manners to his son Herbert, telling him to study well, work hard, grow to be a worthwhile member of society and a credit to his Manners heritage. He does not appear to have taken this advice since, five years after his father died, the letters indicate that Herbert was in some trouble and the family was not happy with him. Manners' letter was signed, "Adieu my Dear Herbert, ever yours affectionately, R. Manners" and was dated "Bloxholm 10th November 1821."

Herbert, now entering his sixteenth year of age, received a long letter from his godparent, which read, in part:

> My dear Godson, I had hoped, after what passed some time ago, and the assurances I received through your Mother [Mary Ann Mansel] that you would – by steadiness, attention to your studies, and submission to your teachers to – the great kindness and indulgence of Lady Robert Manners [his grandmother], Mr Manners [his uncle the Hon. George Manners, his father's youngest brother] and others who are pleased to interest themselves in Your Welfare – no reproach of your Conduct other than satisfactory in showing your determination to do credit to the Memory of your excellent Father to realise his – wishes that you should become an useful and honourable Subject of His Majesty's and a respectable and gentlemanly Member of Society [unreadable] ... you are mistaken if you imagine that you can indemnify yourself for the loss of the good will and approbation of those in whose circle you have *now* the option to move, by resorting to those of inferiority status and more humble habits.
>
> Your sincere friend And Affectionate God Father [Signature indecipherable, but it might be C. Taylor]
>
> 55 Portland Place, London, January 27, 1828

And another from his aunt, the Hon. Mary (Manners) Hamilton Nisbet, the sister of General Robert Manners, written five years after Manners had died:

> In transmitting this letter to you, I will add my prayers and best wishes that it may bring you to a proper sense of your past conduct, and that it will make you endeavour very seriously to [unreadable] – Remember it is from the Dead [meaning she is speaking on behalf of Herbert's now dead father, her brother General Robert Manners]. One who never knew of your conduct or of your having left Mr. Dimma, [Herbert ran away from school – from several of the letters it was clear he had not been happy there] ... Nothing but the deep interest I take in your moral welfare would make me write this to you, as having your sister now – my Charge adds much to the anxiety I feel for your turning out Respectable ... of the care my Dear Brother bestowed on you. [signed] M. Hamilton Nisbet. Portman Square, April 21st, 1828.

The nature of the bad behaviour wasn't revealed, but the suggestion seems to be that Herbert was making friends who were not from the most desirable of backgrounds, and wasn't applying himself to his school work. Indeed, it is clear that at the height of the problems he ran away from school at the Reverend Thomas Dimma's, South Queensferry, Edinburgh.

Charles Childs, the reason for Kathleen Luckcuck's letter to me in 2002, has not yet been mentioned, but I had to set the scene of what was happening to his mother, Mary Ann (Goodchild) Mansel. In all but name she was a married woman with six children. General Manners had set her up with a home in Chelsea, in Park Place South, very close to the King's Road. He does not appear to have lived with her on a permanent basis, since his London home was in Curzon Street, Mayfair. In 1823 he died there. Manners was a General in the army,[399] an Equerry to the king,[400] and a member of parliament. He resigned from parliament in 1820.[401] Naturally, he had to keep his private life separate from his professional duties. Their respective residencies were one and a half miles apart – not far, but far enough for discretion to be maintained. His family seat was at Bloxholm Hall, in Lincolnshire, where he also spent some of his time, and enjoyed having his children visit him there, which is clear from the 21 letters I transcribed.

The Charles Asgill Affair

How did General Sir Charles Asgill come to have a relationship with the mistress of General Robert Manners? That is a question which nobody has the answer to. The closest guess is that since Charles Asgill was Equerry to HRH the Duke of York, and Robert Manners Equerry to His Majesty, King George III, the two men would certainly have known one another and perhaps an innocent introduction was made one day – and the rest followed? Charles Childs was born to Mary Ann before 6 May, in 1816, by which time Charles Asgill had left Dublin and was living in London, and had been for some time already.

Kathleen's findings were based on a book, or rather more accurately, a scrapbook, entitled "Father's Book, Members of the Family" which appears to have been compiled by two of Charles Childs' grandsons, after his death in 1884. Was the secretive Mary Ann outwitted by her son, Charles Childs, leaving a clue for posterity by incorporating her Manners/Mansel surname for some of his children? It is a scrappy scrapbook, written by uneducated people who were trying to make sense of their grandfather's story. Kathleen was a descendant of William Herbert Childs, and I am a descendent of Elizabeth Mary *Mansel* Childs; she was my great-grandmother, who married Frederic Aylin Edmed on 29 October 1878 at The Parish Church, Loose, Maidstone, Kent. *This is the church where Charles Childs and Mary Ann Mansel are buried in the same grave,* thereby bringing my story right into theirs. Elizabeth Mary Mansel Childs was Charles Childs' fifth child, of eleven in all. Another cousin of mine, DNA confirmed, is Leigh Childs, a direct male line descendant of William Herbert Childs, so a closer cousinship with Kathleen than was mine. We've never found a marriage certificate for Charles Childs and the mother of his eleven children, Elizabeth Hayward. They lived as husband and wife for several decades though.

Charles Childs (the son of General Asgill and his shared-mistress, Mary Ann) was born sickly, at the Officers' Quarters, Life Guards, in Whitehall, and nearly died within six weeks of birth. He did survive. His birthplace seems strange to me – he was The General's illegitimate son, according to the family scrapbook, and yet it can only have been Asgill or General Manners who allowed Mary Ann to give birth at the officers' quarters. But how can we know?

Since Mary Ann is believed to be of Irish Roman Catholic stock, she arranged for Charles to be baptised urgently, since he was, according to the scrapbook, "in danger of death" at birth. The ceremony was performed privately at home. She engaged the services of Charles Edouard

Charles Childs

Drummond, 5th Earl and Duke of Melfort and 10th Duke of Perth (1752–1840), titular 13th Earl of Perth, "Abbé de Melfort." The scrapbook states that the Duke of Melfort D.D. "witnessed" the baptism, but since the Duke had a family connection to Peter Drummond-Murray (my one-time boss at Arbuthnot Latham, in the city of London), I asked him if this was a likely scenario. He was adamant that if the Duke was present, as claimed, then he would have performed the baptism himself. Lengthy searches have been undertaken, over so many years, but the baptism entry for Charles Childs has never been found. I very much doubt it would have told the truth, even if it had materialised.

Charles Childs grew up with his half-siblings around him, and seems to have been a bit of a tearaway character from the outset. I tend to think of him as a cuckoo in the nest. On 27 May 1833, Lucy Mansel wrote to her brother, Herbert Mansel:

> You will I hope be able to get leave in the Autumn as George hopes to be at Biel at that time and the Grouse Shooting commences on the 12th. I leave Sutton on Saturday and go to Mrs Ferguson's at Portman Square. I hope George will go with me. Ellen Turner is here. Whenever Charles [aged 17] wishes to keep her in proper awe of him he has only to present an unloaded gun at her head and she screams for the fare life as they say in Ireland. I shall now leave the rest of this paper for Mama.
>
> Your affectionate and attached Sister, Lucy

It seems that Charles Childs grew up with a chip on his shoulder, which he never appears to have shaken off. Perhaps he felt "deprived," and maybe with good reason. Perhaps his mother doted on him too much and spoilt him. Mary Ann's "Mansel" children were well provided for by their Manners father, but other than his mother's love, this does not appear to be the case for Charles. It must have been hard growing up in that environment – the "them" and "us" syndrome. He had five elder half-siblings, and one younger, although he would not have remembered the last, since he had died so young. Whatever it was like for him, clearly his mother loved him and they lived together all her life. Charles was 38 years old when Mary Ann died. Perhaps Mary Ann felt pity for him that his father had not left him well-provided for, as General Manners had done for *his* children. Whatever the case may be, Charles Childs seems to have been an aloof and strict father himself when his time came, and not close to his own children at all. Was he taking General Asgill as a role model in this respect? He made them all bow and curtsey to him on meeting and one of his children ran away from home at the age of 12, and joined the army.

As for Charles himself, he lived in Loose, near Maidstone in Kent, all the while with his mother, and "dabbled" in hop farming.[402] In spite of having half-siblings, he inherited *everything* from Mary Ann, according to her will, including her house. Kathleen's findings indicated this was an impressive house in Loose, "Hill House," but this has not been confirmed with supporting documentation. Family stories say that Charles was a "ne'er-do-well" and lost a small fortune. I've seen evidence that he became bankrupt, but his plea was that he was owed a lot of money by creditors who had not paid their debts. The newspaper notice for "Maidstone: A Farmer and Hopgrower's Bankruptcy: Of Loose and No. 40 Great Dover Street, Southwark," gave the reasons as: "badness of trade; bad debts; losses on growing and the sale of hops; insufficiency of capital and illness of the family."[403]

Hill House, Loose, today. Author's collection.

Was he happy within his 'marriage'? Possibly not. There is no proof of a marriage, but he certainly had a live-in relationship with his housekeeper, Elizabeth Hayward. On the 1881 census she described herself as a widow, which clearly she was not given that he was still alive.

Charles Childs

Why would she do that? Were they separated by 1881? He died in Barming Heath Mental Asylum, on 6 May 1884, and is buried in the same grave at All Saint's Church, Loose, with his mother, Mary Ann. Their badly damaged gravestone reads: "In loving memory of Mary Ann Mansel Of This Parish Departed this life 26 October 1854 aged 71 years, For as in Adam all die, even so in Christ shall all be made alive. It is the spirit that quickeneth. The flesh profiteth nothing. (1 Corinthians c15:v22) Charles Childs of This Parish Who departed this life 6 May 1884 aged 68 years." Much of the inscription is so badly damaged that it is either missing or almost illegible. I have completed the biblical verse for fuller understanding, otherwise it is too disjointed.[404]

Charles' 'wife' Elizabeth Hayward and two of their daughters are on the same plot in that Loose Churchyard. Census returns have Charles as a "visitor" in Mary Ann's house. Keeping her private life private is one thing, but denying her son seems extraordinary to me. He was the *sole executor* of her will, and the *sole beneficiary*. None of Mary Ann's "Mansel" children inherited anything from her. Since his body lies in her grave, he must have been one extraordinary 38-year-stay-"visitor"!

> This is the last Will and Testament of me Mary Ann Mansel of the Parish of Loose in the county of Kent Gentlewoman and whereof I appoint Charles Childs now residing with me sole Executor and direct payment to be made by him of all my just debts and funeral and testamentary expenses as soon as convenient may be after my decease I give devise and bequeath unto the aforesaid Charles Childs my Executor all my Real Estate and Personal Estate of whatsoever nature or kind for his entire use and benefit absolutely and for ever and lastly I revoke all my former Wills Codicils and Testamentary dispositions In witness whereof I have hereunto set my hand this twentieth day of February one thousand eight hundred and fifty four
>
> Signed by the said Mary Ann Mansel the Testator as her last Will and Testament in the presence of us present at the same time who at her request in her presence and the presence of each other have subscribed our names as witnesses Isaac Russell – Harriet Russell
>
> Proved at London 30th April 1855 before the Worshipful ….. ….. …..Doctor of Laws and surrogate by the oath of Charles

Childs the sole Executor to whom authority was granted having been first sworn.[405]

Mary Ann Mansel and her son, Charles Childs, in the same grave together in All Saints' Churchyard, Loose, Kent. Author's collection.

Charles Childs

In my view, the evidence is there that Charles Childs was Mary Ann (Goodchild) Mansel's son. But was he the son of Sir Charles Asgill too? I will come to what evidence there is in the following chapter. There is absolutely no evidence to suggest he was the son of General Robert Manners. However, Mary Ann had her final child with Robert Manners, Henry Edward Mansel (1818–1820), *after* Charles Childs' 1816 birth!

Charles Child's death was tragic and sordid. His medical records, which are written into large bound volumes, with printed headings, follow:

Case notes of the county Asylum, Barming

Friend: Elizabeth Childs (wife) Loose-hill, Maidstone [here she is called his wife, with the surname "Childs", so maybe the marriage did occur, but the certificate has been lost?]

Religion: C of E [Church of England, another oddity since he was baptised by a Roman Catholic priest]

Suicidal or dangerous: Both

Clean or Dirty – Dirty

Form of mental disease – Dementia Senile.

History on admission – Is silent – refusing to eat or speak, and also refuses to be undressed – Is in a very depressed state of mind.

Elizabeth Childs: Wont do anything he is told: messes himself in his bed: is violent.

Accompanied to Asylum by Mr Treasure, Relieving Officer.

Admitted to Ward no 4 – A H Hillier [illegible]

On admission – [no date of admission is given, so his length of stay is unknown] Patient was admitted in a quiet demented state, appeared to imagine he was going to be hurt, so resisted having his clothes taken off – would not speak a word – ideas are apparently very confused and chaotic, memory appears almost lost.

Physically, he is a little [his father, Charles Asgill's nickname was 'Little Charlie'], poorly nourished old man, expression vacant and melancholic, chest narrow with sunken intercostals, vesicular manner loud and harsh, heart sounds are indistinct, appetite is bad, tongue coated.

March 18, 1884 – was rather restless through the night: will not speak at all: looks about him in a frightened, nervous manner, as if afraid of harm.

March 20 – Is a little quieter now – appeared to think he was going to be injured in some way: seldom speaks: takes nourishment better than formerly.

March 25 – Appears more settled and contented here: is quite lost and demented: cannot converse at all intelligibly but understands simple questions fairly well.

April 10 – Quite demented and very feeble – can only get about with difficulty: is not so restless, but still very melancholic: appetite is pretty fair.

April 28 – Keeps in much the same state – sleeps but little - never speaks unless first addressed: appetite keeps pretty fair – he is very weakly still however.

May 5 – Has fallen off a good deal in appetite: is on extra diet: ordered to bed: have written to friends to say he is not so well.

May 6 – Became suddenly much worse early this morning and died at 6:15 am, in the presence of Charles Hillier, C.G. [perhaps C.G. stands for caregiver?[406]]

May 7 – Post mortem made this day and copy of following notice sent by Messenger to Coroner for Maidstone –

Charles Childs died on the 6th day of May 1884 of Valvular Disease of the Heart, the duration of the disease being several years.

F Pritchard Davies MD, Superintendent   Coroner for Maidstone
Dated May 7

Charles was 64 years old when he died, and had a sad and sordid end in a Victorian lunatic asylum; an end to what appears to have been a sad and unfulfilled life.[407] It seems to me that he expected everything to be handed to him, without putting in the graft. Perhaps he considered that had been the case for his half-siblings? All Mary Ann could do to redress the balance was to leave all her worldly goods to him, and welcome him into her own grave.

CHAPTER TWENTY-TWO
Mary Ann (Goodchild) Mansel

It is clear cut that Mary Ann was the mistress of General Robert Manners, over a long period of time. She is mentioned in his will, leaving her an annuity for life, along with financial provision for their children.[408] Much harder to verify is the evidence in "Father's Book, Members of the Family" which states that she was also the mistress of General Sir Charles Asgill and gave birth to their son, Charles Childs. I never dreamed that the missing proof would come in the form of an unannounced (and therefore unexpected) DNA test, the results of which came through on 3 May 2021. A descendant of General Manners, a New Zealander by the name of Dean Crowley, had a DNA match with me, putting us as 3rd-5th cousins on Family Tree DNA's website. He is actually a 4th cousin once removed. Since this match I have now matched with descendants of other surviving Mansel children, which makes my DNA match to Mary Ann's Manners children conclusive. By contrast, a DNA match with a male line descendant of Charles Childs, Leigh Childs, puts us as also being 3rd-5th cousins; whereas he is actually a 3rd cousin. The purists involved with this story never accepted that Mary Ann really was the mistress of two generals. But DNA has provided evidence, which had never been found in the archives. At the time of writing, the DNA and genealogy is being worked on by Meg Bower, a Family Historian based close to William Charles Asgill's stamping ground of Beddington, in Surrey. It was Meg I contacted in March 2015, when Andrew and I were searching for WCA's birth in Ewell, Surrey.

Believed to have been born in Waterford, Ireland, Mary Ann Goodchild (before 26 October 1780 – 26 October 1854) made her life in London and was soon settled in a long-standing relationship with Manners. He died on 9 June 1823 at his house in Curzon Street, and was buried with his parents at the church of St. Mary the Virgin at Bloxholm.[409]

In my search for Mary Ann, I was delighted to come across another court case. It provided me with her address, which became very useful when connecting her up with Sir Charles Asgill. On 9 September 1818 the trial of Richard Wheeler took place at the Old Bailey.[410] Wheeler had stolen a hog from Mary Ann, and although he returned the hog, he was deported to Australia for 7 years. The trial proceedings revealed Mary

Ann's name and address in Chelsea as 15 Park Place South, very close to the Man in the Moon public house.

Mary Ann Mansel's Sun Fire Insurance cover increased substantially in 1821 (just 4 months after she had already renewed it).[411] The timing coincided with Asgill selling his York Street house (after his wife Sophia died in 1819). She may have needed extra cover if Asgill moved in with her after selling his house. Two codicils to Asgill's will were written from her house in the days before his death.

His obituary in *The Gentleman's Magazine* noticeably made no mention whatsoever of his place of death.[412] This is very uncommon for a notable person. Manners' obituary stated he had died at "his house in Curzon Street" and Sophia's that her death had occurred at "her house in York Street." It would seem discretion was the better part of valour with regard to Asgill, and mention that he had died at his mistress' house would not have been *de rigueur*.

When both Mary Ann's lovers died in the summer of 1823 she fell to pieces, and her surviving four Mansel children became wards of Manners' brother George (1763–1828) and sister Mary Hamilton Nisbet (1756–1834). Another court case mentions the wardship, the late General Robert Manners, Mary Ann Mansel (who had to give evidence), her surviving Mansel children, and also Robert Manners' sister Mary and brother George. It appears they were fighting over who got what from Robert Manners' will.[413]

Of all her surviving five children (in 1823), Mary Ann only kept Charles Childs with her before she eventually left London. She firstly moved to Sutton Valence, as attested in Lucy Mansel's letter to her brother Herbert,[414] then to Lenham, also in Kent. She finally moved to Loose (near Maidstone in Kent, where I was born). She seems to have acquired Hill House in Loose, according to "Fathers Book, Members of the Family," but archival evidence has not been found. She is on the 1851 census as being at Loose Hill, Maidstone, Kent with Charles Childs, his "wife" Elizabeth Hayward and four of their children.[415] Astonishingly Charles Childs is listed as a "visitor," although he lived with her his entire life. Why did she have so many members of his family living with her, if he was not her son?

When her Mansel children became wards of the Manners family it split the family permanently and the younger ones, in particular, became nomadic, spending their time either at boarding school, at Manners family homes in London, or at Bloxholm Hall in Lincolnshire. They never

Mary Ann (Goodchild) Mansel

became a tight family unit again although they wrote to one another often and always went to visit Mary Ann as frequently as they could. While Herbert Mansel caused his family much worry with regard to his behaviour, they all ended up having successful lives and careers. It is believed that Mary Ann's move from London to Kent was to be near one of her Mansel sons, by then in the army at Invicta Park Barracks in Maidstone, a depot for the British Cavalry.[416]

After Mary Ann Mansel died, on 26 October 1854, her death certificate recorded her as being the "widow of Robert Mansel, Gentleman," a ploy used to continue to hide her tracks. They were not married, so she was not a widow, and her partner's name was Manners not Mansel. Were it not for surviving family letters which make the real situation known, her secrets might not have been revealed. Her extreme privacy has made her extraordinarily difficult to find. She was buried at All Saint's Church, Loose and 30 years later, in 1884, Charles Childs was buried in her grave.[417] I think this clinches the mother-son relationship.

Mary Ann Mansel's Death Certificate. She died on 26 October 1854 (Maidstone, Volume 2a, Page 269). One of the few verified documents which proves her existence. Public domain.

It is curious to speculate as to why General Sir Charles Asgill wrote codicils to his will from Mary Ann's Park Place South address, where he was living for health reasons, or so he said, and why her Sun Fire

The Charles Asgill Affair

Insurance was increased at the same time as Asgill sold his York Street house. The first codicil, written on 15 July 1823, began, "I, Charles Asgill, General in His Majesty's Service, now residing here for my health, add this a Codicil to my last will …" The second, written the next day, included:

> To my friend Mary Ann Goodchild, otherwise Mansel, the white clock on the mantelpiece with the two large vases; the black mare with saddle and bridle to her son William Herbert Mansel. To Sir Charles Ogle my other horse, carriage and gig. [Signed] Charles Asgill /-/. Witness W.G. Maton MD of Spring Gardens"[418]

William George Maton M.D. (31 January 1774 – 30 March 1835), English physician, who treated Asgill when he was dying, and witnessed his Codicils. Artist unknown. Public domain.

Mary Ann (Goodchild) Mansel

What on earth were Asgill's possessions doing on her mantelpiece, if not favourites of his, brought with him when he moved in? This isn't much to leave to a woman who bore his son, but the terms of his father's will were such that his hands were tied in the event he died without legitimate heirs. Besides, General Manners had left her well provided for.

Why, also, did he leave his black mare and saddle, not to Charles Childs, but rather to that naughty boy, Herbert?! Perhaps it was because Charles was only 7 years old when Asgill died. Or perhaps he decided not to acknowledge his son, just as Mary Ann did not acknowledge him as her son, either.

These codicils show that Mary Ann's house is where he died a week later, on 23 July 1823. He stated that he was living there "for my health." But for how long had this been his reason; or his excuse? Mentioning that his clock was on her mantelpiece is a very intimate detail too. Surely this means that his remaining possessions were located in her house? Since Mary Ann called her son, Charles Childs, a "visitor," although buried in the same grave with her, Charles Asgill calling Mary Ann his "friend" is really a step up from her approach. After all, Robert Manners called his son, Herbert, his "friend" as well. Perhaps this was common usage in the era, especially if you wanted to hide the real relationship? Add to that Asgill's obituary in the *Gentleman's Magazine* giving no indication of his place of death, which is so unusual. What other conclusions can be drawn but that "Father's Book, Members of the Family" has got it right?

I no longer expect to get any closer to the truth, so I must conclude that Charles Childs' parentage was as explained by his grandsons in their scrapbook. Others may find evidence which tells a different story; but that will have to be another book, another day.

CHAPTER TWENTY-THREE

The Journey of Discovery

Anyone who has dabbled in genealogical research will know that the "ups" come with the "downs." One minute one is elated at unlocking a previously locked door – the next minute a door slams in your face. Blocked; finished; done! It goes with the territory, and certainly has many times in my case. But I'll say right from the start, I wouldn't have missed a minute of the past two decades. It has been the most incredible journey imaginable. I have worked insane hours on this project, on a 24/7/365 basis for 21 years, clocking up far more hours than I ever did during my seventeen years in paid employment. It has also cost me thirty-five thousand pounds, which looks less in words than figures! And it hasn't stopped yet.

The best part of it has been the cross-section of people I have made contact with. From the top of the academic spectrum there were several professors in two continents, through to many descendants of this story, taking me from Australia, New Zealand, America, Europe, and Africa, (where I traced the same family found by Katherine Mayo), and Scotland (where I traced the same Gordon family found by her too). They have all played their part in revealing the truth, where truth has been possible to find. There's more to do; but I doubt if this journey will end in my lifetime.

It will come as no surprise to learn that finding Asgill's hidden letter to the editor of the *New Haven Gazette* was the pinnacle of pinnacles in terms of elation. I'm living in my sixth house since this began in 2002, and several were overseas, so Google has definitely been my friend. It was daunting to begin with, because there were just too many "hits" returned by a given search, and I often found that the Asgill names revealed were not the ones I was particularly interested in. Charles Asgill's father, the 1st Baronet, and their kinsman, Translated Asgill, kept popping up too and I found myself spending too much time looking at entries which I had, by then, already become as aware of as I needed for my purpose.

So, one day I decided to do something I had never done before: I started sifting through the search results from the very last page, rather than from the first. I was going through too much "old ground" by beginning at the beginning. I had moved back towards the beginning, about fourteen pages, when I saw the "For Sale" notice for Asgill's letter. I could not believe my eyes. My heart was beating so fast I thought it

The Journey of Discovery

would jump out of my chest. I knew I could not afford the asking price (back then, the $ conversion was £10,000), so Professor Robert Tombs at Cambridge university stepped in, acquiring a photocopy. He is most definitely my hero. I wanted to contact all the Ivy League universities in America, to see if there was interest in buying the letter, so that it could be in the public domain, but having started with Yale in New Haven, it took too long (three months) for them to give me their negative answer, by which time the item had been sold.

My research trip to America (13 to 30 May 2019) was the turning point, with three significant outcomes. The first leg was Huddy country. Then came the bonus of meeting Professor Gregory Urwin, in Doylestown, Pennsylvania, who had been so willing to answer my questions, over the years. Lancaster resulted in Asgill's letter being published after 233 years; Summit (Chatham, in Asgill's day) resulted in learning about Timothy Day's Tavern, which I am convinced was the location of his close confinement; and my visit to Trinity Church in New York City resulted in a stanchion honouring Lieutenant Colonel James Gordon. It was a rewarding trip, which also introduced me to some lovely people.

The reason I travelled to America was to discover where Asgill had been held as a close prisoner. History books tell us only "Chatham, New Jersey" with some adding that "Washington ordered him to be sent to the Jersey Lines, under close arrest." But these history books do not relay that Asgill himself wrote that he had been held in a "tavern" in Chatham. So my task, when I reached the area of Chatham now known as Summit, was to discover what that tavern had been. With the guidance of Robin Carroll-Man, at the Summit Free Public Library, I discovered that there were only two taverns there in 1782. One was Timothy Day's Tavern (located nearby to Colonel Elias Dayton's home) and the other was to the south of the then village. The former was a staging post, used by travellers passing through on their way to or from New York City and beyond. It was even visited by George Washington himself. This tavern had beds for weary travellers to sleep in, and provided food for their guests. It even operated as a temporary hospital for wounded patriots from time to time. And the French provided the landlord with ovens so that he could bake bread for the patriot army. In other words, Timothy Day had the wherewithal to house a prisoner two. But was it the other tavern, owned by Sylvanus Seeley? He had a tavern/store in Chatham that was reportedly very successful. Seeley was a Colonel in the New Jersey Militia and

The Charles Asgill Affair

Dayton was a Colonel in the New Jersey Regiment. It seems Seeley was known for the quality of his liquor, which may explain Asgill's problems with drunks. While Day's would have been closer to Dayton's, Seeley's was not much further. My view is that Seeley's "Store" (which happened to have a bar in order to provide liquor), is much less likely to have had overnight accommodation. Regrettably, I cannot prove my theory, but I am firmly of the opinion that Timothy Day hosted Charles Asgill. His was "the tavern with the mostest" – attracting not only George Washington,[419] but the French too.

One day I was wondering if I would ever know anything about Asgill's house in York Street, London, since I had been unable to find it when I went there to visit his stamping ground, and to go to St. James's Church, Piccadilly, where Sophia and Charles Asgill are buried. A casual enquiry of an archivist, Liz Brown (at the time with English Heritage) resulted in me being sent a lot of information about the house. This was over and above my original request. I was so touched, but she explained that she had been doing some research herself, having been interested in Asgill's story as told to her when I phoned for the information I was then seeking. What a kind gesture, which I have never forgotten.

St. James's Church was one of those brick walls, or closed doors, since there are no records of memorials to either Charles or Sophia Asgill there, although their burials are recorded in the parish records. Nothing has ever turned up on this score through Memorial listings either. I always found it very strange that Charles Asgill never commissioned a plaque to commemorate his wife, whom he described as "my beloved wife" in his will.

Having left no legitimate heirs behind, there was no one but his two surviving sisters, or members of the Ogle family, to do the same for him. But that too, apparently, never happened. I did track down two construction workers who had worked for Rattee & Kett, Cambridge, who had been the contractors to restore the Christopher Wren church after it was bombed in the Blitz of London on 14 October 1940. They remembered that under-floor-heating had been installed at the church, which had entailed removing, temporarily, all the coffins in the crypt. That was interesting, because it must have involved the two Asgill coffins. The mystery of no plaques is unlikely to ever be solved. I have dreamed of it being possible to remove Charles Asgill's coffin again, and take a DNA sample from his remains! I am, as I write, in touch with the Reverend Lucy

The Journey of Discovery

Winkett, at St. James's Church, Piccadilly, in the hope that I may one day be allowed to erect a plaque to Charles and Sophia Asgill.

Finding Sophia Ogle's portrait was one of the greatest highlights. I was overseas when my daughter, Nina, managed to find information about the Hoppner portrait of Frederica, Duchess of York, which simply stated that her Ladies in Waiting were also in the image (three of them, so which one, I asked myself). Then I had help from Deborah Gage, who sent me a picture of Sophia as a child, cuddling one of her Red Setter dogs. It was a clear match to her adult image with the Duchess. This was another jigsaw puzzle, but the pieces did slot together. Finding where the Hoppner image is located now was a far greater task though, since records of fine art in private collections are not easy to come by. I was eventually allowed to view it (in the mansion of a Saudi prince) and take photographs. Her image was larger than life-size, so I came away feeling I had met her.

I have always had a burning desire to see the Thomas Phillips portrait of Charles Asgill, painted in 1822 and exhibited at the Royal Academy that year. I have been on the lookout, contacting so many galleries and art experts over the years, but its whereabouts is one of the as yet unsolved mysteries. Engaging the services of a firm of stolen-art specialists in Washington, DC brought me not one jot closer to a resolution, although it dug deep into my pocket. I am fortunate, though, to have one of only 100 mezzotints of the Thomas Phillips portrait. Free of charge too, from someone who had it hidden in their attic and wanted it out of their way!

I've visited all the main locations, such as Asgill House, Duke of York Street in London (a name-change from York Street) and St. James's Church, Piccadilly several times, and that is about all there is to see where it is known Asgill spent time himself. I've toured the City of London, viewing the Golden Coach commissioned by Asgill's father at the Museum of London; walked Lombard Street where the Lord Mayor's bank was; and at least seen the Mansion House from the street, even if I didn't go inside. Some of my visits to the City of London were achieved when I was invited to the January 2012 social event organised by Paul Lay, the Editor of *History Today*, for contributors to their magazine the year before. Several distinguished historians were present, some of whom I recognised from having seen them on television. I got talking to a few who were in the film business and, I thought, had managed to interest at least one to take the matter further. But it never happened.

The Charles Asgill Affair

But I did have some contact with Julian Fellowes, who I would have dearly loved to write the screenplay. I approached him when, unknown to me, he was heavily involved in writing the first series of *Downton Abbey*, so he was too busy, but professed some initial interest. Longer lasting communications pursued between his wife and me, but it sadly led nowhere. Many people in the film world have been contacted, and I even had a contract with a producer in Hollywood, who said he was very interested in making the film, if he could get the funding inside of one year, when my contract would expire. Again, it came to nought. Now there is Professor Peter Henriques, whose 2020 book *First and Always: A New Portrait of George Washington* included a chapter dedicated to The Asgill Affair.[420] He has said how interested he would be in getting a film or television series off the ground. If the film is ever made, I hope I'm still alive and can attend the opening night in Leicester Square!

It always distressed me that James Gordon had not only been buried in an unmarked grave at Trinity Church, New York City, but that his burial entry in the church records had been so carelessly entered, with his rank abbreviated to "L.C." and his surname misspelt as "Gorden." So, when I travelled to Lancaster to try to get Asgill's letter published, our plan was to end our journey in New York City. There, we had the great pleasure of meeting the Rev. Kristin Miles to explain that James was such a special person, offering to lay down his life for a younger man, that I had brought with me a little plaque, and would they install it somewhere in the church? After returning to England I heard that my plaque did not conform to the church's house-style as it were, but that they would install a stanchion in their graveyard in his name. I was elated. But delays occurred; firstly because of renovation work on the church building, and then the Covid-19 pandemic. I kept in touch with Kristin, to see how things were progressing, and she kept me abreast of the situation. I was overjoyed to hear that this stanchion was erected on 8 March 2022. James Gordon now has the memorial he so deserves. My plaque was returned and I gave it to Robert Balfour, the Lord Lieutenant of Fife, who is James Gordon's 4x great nephew. James himself never married nor had children of his own.

One of my greatest disappointments was being unable to acquire a silver hot water urn presented to Asgill by the grateful loyalists of Clonmel, Tipperary, after the Irish uprising there in 1801. It is really beautiful, but I found the auctioneers listing only after the sale had gone through. I tracked down the new owner, the now late Professor Bill Benson, in Canada, and asked if he would sell it on to me, at a profit. He

The Journey of Discovery

refused my generous offer! This happened again, when I found an auctioneer's listing for a snuff box presented to Asgill by the grateful loyalists of Kilkenny after the Irish Rebellion of 1798. Once more I tracked down the new owner, with the same disappointing results. No amount of profit would tempt him to sell it on to me! How I would love to own something once owned by Asgill. I did manage to purchase a drawing of one of Sophia's Red Setter dogs, which had been created at Lynedoch's estate in Scotland, by Charles Loraine Smith (1751–1835). It is hanging on the opposite wall to the desk from which I am writing now.

Not finding Charles Childs' baptism certificate was desperately frustrating, but would it have shown his parentage, even if it had been found? That's another unanswered question. Although not quite so important, not finding his marriage certificate either was a disappointment. There were other brick walls in this journey of mine, but I've mentioned the most important ones.

The identity of the "Colonel Gregory" assisted by Asgill in 1781, has actually been the hardest mystery of all. I am grateful to so many Americans, for their very kind help in trying to get to the bottom of the matter. Latterly Chris Noel joined the team and nobody could have shown more enthusiasm than he to solve the mystery; he simply could not have been kinder or more helpful. Alison Thurman (North Carolina Archives) and the historian James Taylor put an enormous amount of effort into the project and I am so grateful to both. Sadly, the mystery was never solved, further than reaching the conclusion that there is sufficient evidence to suggest that the incident *could* certainly have happened.

The following article is about Colonel Robert B. Hicks Jnr. (1759-4 September 1829). He has already been mentioned, as the man who buried a pot of gold which was never retrieved. He was almost the same age as Asgill, who was my British 3x great grandfather; Hicks was my American 4x great grandfather. Both Whigs, they fought on opposing sides during the American War of Independence. Had Asgill arrived in America sooner, then the two would almost certainly have met on the battlefield of Guildford Court House (15 March 1781) in North Carolina, where Robert distinguished himself by fighting the British, single-handedly, when the Patriots fled the field of battle. Robert Hicks was in the North Carolina militia, whose Brigade Commander was Brigadier General Isaac Gregory. The big unsolved mystery is whether all three did fight on the same soil at the battle of Black Swamp (26 June 1781). It is certainly possible.

The Charles Asgill Affair

The Hicks family were distinguished amongst those worthy of remembrance in Granville. Captain Robert Hicks lived about a mile from Oxford in 1770. Robert Hicks entered the revolutionary army, and was in the battle of Guildford with the North Carolina militia where these raw and undisciplined troops were placed by General Greene in the front line, and there, overwhelmed by the British, fled; young Hicks stood his ground, and fought single handed, until nearly surrounded, and after his men had gone a considerable distance, he then escaped and shared, during the remainder of the war, its dangers and its glories. The family is English, and settled in Brooklyn, New York, in the locality now known as Hicks Street. The family was distinguished in England for its courage and ability, and one of them was knighted for his deeds of daring."[421]

Captain (later Colonel) Robert Hicks was 21 years old at the Battle of Guildford Court House. I am every bit as proud of being his descendant, as I am also that of Charles Asgill.

Research can be fun, and it can be anything but! I have discovered that even eminent historians out-and-out lie. The following is something which gave me a huge amount of trouble and, at the time, much anxiety. I was badly stung, not for the first or last time.

When I saw the following quote, in *The Life of George Washington, Studied Anew* by Edward E. Hale (1822-1909), a revered American historian, I was convinced I had found the missing pieces of the story of Asgill's experiences in America – his debriefing in London in 1783. Something I always felt should have been there, but I had never found. I engaged the help of everyone from the Library of Congress, the British Library, professors in the US and UK, and Uncle Tom Cobley and all. Dozens of people worked on this for about six months, so important a document did I believe it to be. Hale wrote:

> Here is a curious note of Asgill's, which we print from his autograph, written in London the next year; "In answer to your question if the Americans put me in irons during the time of my confinement, for their sake as well as mine, I have the satisfaction to inform you that they never did. Henry Asgill."[422]

I suppose Hale thought he could get away with this by giving Asgill the name "Henry." The following review of Hale's work was found by Meg Bower: "It must be admitted, however, that Dr. Hale has rather a curious idea of the value of tradition. He repeats a story of George Washington which he frankly admits he made up out of his own head, or, as he expresses it "I did try my hand on starting a tradition ... Now, there are many stories in Plutarch which have no more foundation than this. There is no proof that this is false, so let us hope it is true."[423]

The Harvard Square Library comments thus, after praising Hale for his unbounded popularity, good works and philanthropy:

> His historical writings were animated and picturesque, but he cared more for the general sweep of events than for accuracy in detail. He made history interesting but not always precise as to facts. He believed in original sources, but he was in too much of a hurry to seek for them.[424]

What I am saying is that history is littered with misinformation, the majority of it (Hale excepted) created accidentally. For instance, there are two items of incorrect information in my *History Today* article,[425] which I most certainly did not know at the time of writing, and which, now, make my toes curl. One involved a vicar putting out very misleading information, attributing his own imaginary words to, as it seemed, a sermon given by Joshua Huddy's commanding officer, Colonel Asher Holmes, and managing to make it sound like fact. I fell for this misinformation hook, line and sinker. The other involved one of the Lord Mayor's descendants believing a family hand-me-down story, which turned out to be inaccurate, causing me to believe this misinformation. We who dabble in history, are very likely in for some rude shocks, and I certainly found Hale claiming to be writing Asgill's own words, really shocking. Whether Hale ever admitted this untruth at a later date, is unknown.

Throughout the pages of this book I have endeavoured not to take a leaf out of Hale's! And it is a relief to have finally set the record straight on many levels. It is also hoped that others will take this story to the next level and that they will be able to knock down the remaining brick walls; and while I have told Asgill's story to the best of my ability, on the evidence so far found, what, for instance, happened to Asgill's Thomas Phillips' portrait? Was it burnt in a fire in Hampshire in the early 1900s, as has been suggested to me? How did Asgill's Letterbook find its way back to London from New Haven, and where did it go once there? Was it ever returned to Asgill himself? Where, when, why and by whom were Asgill's 1786 attachments removed from the Letterbook? Did Asgill go to the aid of Colonel John Gregory, as I believe he did? Did this happen at the Battle of Black Swamp? Who *was* Colonel John Gregory? Why did Asgill instigate a court-martial against William McCurdy? This I would be intrigued to know, especially since Washington was taking it very seriously, initially. Who was the secretive Mary Ann Mansel? Why did Asgill purloin another man's mistress? What kicked that one off? A drunken night in the Equerry's quarters at Buckingham House (later, Palace)? Was Sophia Asgill Thomas Graham's lover at Woburn Abbey, as I believe she was? What exactly was James Gordon's plan to facilitate Asgill's escape? Did he really tell no one, not even his female accomplices? And in what way were the local ladies going to help bring it about? If any descendants of those women have any information, it is time to share! Why on earth was Caroline (Asgill) Legge's lunacy far too shocking to publish? Charles Childs' dementia, sordid though it was, was preserved in the records. I hope some of these answers will find their way to me before I meet all these people 'on the other side' and can ask them myself! If, dear reader, you can unravel any of them, my email address is anne.ammundsen@aol.com, so do be in touch.

APPENDIX ONE

Letters Omitted from Washington's Published Correspondence on Asgill's Confinement

When *The New Haven Gazette and Connecticut Magazine* published a collection of correspondence relating to the Asgill Affair on 16 November 1786 under the headline "The Conduct of GENERAL WASHINGTON, respecting the Confinement of Capt. Asgill, placed in its true Point of Light," one letter from George Washington to Moses Hazen was omitted, as well as three from Charles Asgill to Washington. These letters put the affair in a significantly different light:

Head Quarters 18th May 1782

sir

It was much my Wish to have taken for the purpose of Retaliation, an Officer who was an unconditional prisoner of War—I am just informed by the Secty at War, that no one of that Description is in our power—I am therefore under the disagreeable necessity to Direct, that you imediately select, in the Manner before presented, from among all the British Captains who are prisoners either under Capitulation or Convention One, who is to be sent on as soon as possible, under the Regulations & Restrictions containd in my former Instructions to you. I am sir Your &c.

G.W. [426]

From Charles Asgill to George Washington, 30 May 1782 [427]

Philadelphia May 30th 1782

Sir,

As I conceive myself under the Protection of a Treaty in which the Honor & Faith of Nations are the Pledges, I have nothing to apprehend but from Hasty Resolves. I must therefore trouble your Excellency with those reasons that induce me to wish my final determination may be deferred until Sr. Guy Carleton can be thoroughly informed of the circumstances of my Confinement. From the Orders your Excellency sent to Genl. Hazen it appears that a British officer being an unconditional Prisoner with the Rank of Captain or Lieutenant, was to be delivered up, that he might be retaliated with for the Death of Captain Huddy, that if no Officer under that Description could be found, this Order then extended to the Captains (British) of Ld. Cornwallis's Capitulated Army - in consequence Lots were drawn for those Captains who were present of that Army & the decision fell upon me. Perfectly innocent of Captain Huddy's Death, & even to this moment uninformed of the circumstances & ever having acted consistently with the Tenor of my Parole I am certain in Justice his Death can never effect me, nor do I know why my Life should be an Atonement for the Misdemeanours of others. I claim protection under the 14th Article of the Capitulation & from your Excellency's known Character I have every Right & Reason to expect it. The same motives that prevailed with your Excellency to require an Officer who was not under the Sanction of a treaty of Faith, will I hope once more induce you to enquire if there are no such Officers at this time of that Denomination, unconditional Prisoners. I shall at present trouble you with no further representations; what other Arguments I may have to urge in my favor are such self evident truths as require no Elucidation. To your Excellency I again make my appeal for Justice & repeat my request that no sudden or hasty proceedings may be held against me.

    I have the Honor to be,

    your Excellency's Most Obedt. Humble. Servant

    Charles Asgill, Lieut 1st Regt Foot Guards[428]

To George Washington from Charles Asgill, 27 September 1782

Chatham Sepr 27th 1782

    Sir

    I hope my unfortunate situation will plead my excuse for being so Sollicotais to obtain your Excellency's permission to return to Europe— as I am by your Command admitted on Parole I am naturaly induced to suppose the motives for my late confinement are removed, therefore let me entreat your Excellency to give me leave to revisit my Friends in

Appendix One

Europe, whose concern for my Misfortune & anxiety for my return, is beyond all-power of Description. Thro Long Suspense my health is greatly impaired, & unless your Excy will be pleased to indulge me in this request, or cause me events to be assured of my fate I fear fatal consequences may attend much longer delay. I hope when your Exy considers that I am not in the situation of a Culprit, that while on Parole I never acted contrary to the Tenor of it that my Chief motives for being so eager for further Enlargment is on account of my Family, these facts, I hope, will operate with your Excellency, to reflect on my unhappy Case, & to relieve me from a state, which those only can form any Judgment of, who have experienced the Horrors Attending it.

I have the Honor to be Your Excellys Most Obedt Most Humbe Servt

    Charles Asgill,

    Lt & Capt. 1st Regt Foot Guards. [429]

To George Washington from Charles Asgill, 18 October 1782

Sir

I have been honored with your Excellys Letter & am exceedingly Obliged by the attention which mine received. I will not intrude on your time by repetitions of my Distress, which has lately been increased by accounts that my Father is on his Death Bed. I have only to intreat as it may be a long while ere Congress finaly [sic] determine, that your Excellency will be pleased to allow me to go to New York on Parole & to return in case my reappearance should hereafter be deemed necessary—if this request cannot be granted I hope your Excellency will give orders that my Parole may be withdrawn, as that Indulgence without a prospect of further Enlargment [sic] affords me not the least satisfaction, I had rather indure the most severe confinement, than suffer my Friends to remain as at present deceived, fancying ever since my first admission on Parole, that I was entirely liberated & no longer the Object of retaliation-- if your Excelly could form an Idea of my sufferings I am convinced the trouble I give would be excused.

I have the Honor to be your Excellencys Most Obedt Hbe Servt

    Charles Asgill Lt & Capt. 1st Guards[430]

APPENDIX TWO

The Conduct of General Washington, respecting the Confinement of Capt. Asgill, placed in its true Point of Light

The New-Haven Gazette, and the Connecticut Magazine[431]

(Vol. I.) Thursday, November 16, M.DCC.LXXXVI (No. 40.)

NON SIBI SED TOTO GENITOS SE CREDERE MUNDO

NEW-HAVEN: Printed and Published by MEIGS & DANA in Chapel-Street.

Price *Nine Shillings* per Annum.

The Conduct of GENERAL WASHINGTON, respecting the Confinement of Capt. Asgill, placed in its true Point of Light.

Messrs Meigs and Dana.

When I was in England, last winter, I heard suggestions that the treatment Capt. Asgill experienced, during his confinement, was unnecessarily rigorous, and as such reflected discredit on the Americans. Having myself belonged to the family of the commander in chief, at that period, and having been acquainted with the minutest circumstance relative to that unpleasant affair; I had no hesitation in utterly denying that there was a particle of veracity in those illiberal suggestions. On my return to Mount Vernon, this summer, I mentioned the subject to Gen. WASHINGTON. He shewed me a communication from London, addressed to Col. Tilghman, which, arriving just after the death of that most amiable character, had been forwarded by his father to the General – by the latter I was also indulged with a sight of his answer. I desired to be permitted to take copies of these papers, together with transcripts from all the original letters and orders respecting Capt. Asgill. Of these I am now possessed.

Anxious that the circulation of truth should be co-extensive with the falsehoods, which have been assiduously propagated; and desirous that the facts may be placed in a true point of view before the eyes of the present age and even of posterity; I have determined, without consulting any one, to charge myself with their publication. It is for this purpose, I request you to insert, in your judiciously conducted paper, the enclosed

Appendix Two

documents, for the authenticity of which I hold myself responsible to the public.

>I am, Gentlemen,
>
>Your most obedient and
>
>Most humble servant
>
>D[avid] HUMPHREYS
>
>New-Haven, Nov. 6th. 1786.

No. I.

Postscript to a letter from James Tilghman Esq. to His Excellency, General WASHINGTON, dated Baltimore, May 26, 1786.

"P. S. A letter is just come to hand from an American in London, who was the friend of my son, in which your name is mentioned, and I take the liberty of inclosing to you a copy of the paragraph. If you think it worth your while to say anything upon the subject, I will transmit it to the gentleman, who writes the letter with some degree of anxiety. I know what pleasure my poor son would have taken in setting the matter in its proper light."

"JAMES TILGHMAN"

No. II.

Copy of the Paragraph.

"I have had it in contemplation to write to you for some time past on a subject in which I find myself more and more interested: I have endeavoured to shake it off from my mind, because I am persuaded that General WASHINGTON is too great in himself to be concerned at any calumny, and his character too fair and pure to need any defence of mine. I have the honour to be introduced to a party of sages, who meet regularly at a coffeehouse, where they discuss politics, or subjects to communicate useful knowledge. This set of men often mention our great and good General, and commonly in a proper manner; but some give credit to a charge exhibited against him, by young *Asgill*, of illiberal treatment and cruelty towards himself. He alledges that a gibbet was erected before his prison window, and often pointed to, in an insulting manner, as good, and proper for him to atone for *Huddy*'s death; and many other insults, all of which he believes were countenanced by General WASHINGTON, who was well inclined to execute the sentence on him, but was restrained by the French General ROCHAMBEAU. I have contended that it was entirely owing to the humane procrastination of our General, that Captain *Asgill* did not suffer the fate allotted him, and that it was most happy to General WASHINGTON's good disposition that the French Court interposed, so as to enable him to save *Asgill*, and at the same time keep an army in temper. This affair is stated by young *Asgill*, and canvassed at the British Court as before related. Now Sir, not for General WASHINGTON's sake, who, as I observed before, is above it, but for mine who take pride in him, as I believe every honest American must, I request the favour that you would inform me fully on the subject, that I may be enabled to parry the only bad thrust made at our hero in my presence."

No. III

Extract of a letter from his Excellency *General Washington*, to *James Tilghman, Esq.* in answer to the foregoing, dated *Mount Vernon, June 5th,* 1786.

"As your son's correspondence with the Committee of New-York is not connected with any transactions of mine, so consequently, it is not necessary that the papers to which you allude, should compose part of my public documents; but if they stand single, as they exhibit a trait of his public character, and like all the rest of his transactions, will, I am persuaded, do honor to his understanding and probity, it may be desirable in this point of view, to keep them alive by mixing them with mine, which undoubtedly will claim the attention of the historian; who, if I am not mistaken, will, upon an inspection of them, discover the illiberal ground in which the charge, mentioned in the extract of the letter you did me the honor to inclose me, is founded. That a calumny of this kind had been reported, I knew: – I had laid my account for the calumnies of anonymous scribblers, but I never had conceived before, that such an one as is related, could have originated with, or met the countenance of Captain *Asgill*; whose situation often filled me with the keenest anguish. – I felt for him on many accounts, and not the least, when viewing him as a man of honor

Appendix Two

and sentiment, I considered how unfortunate it was for him, that a wretch, who possessed neither, should be the means of causing in him a single pang, or disagreeable sensation. My favourable opinion of him however is forfeited, if, being acquainted with these reports, he did not immediately contradict them. That I could not have given countenance to the insults, which *he says*, were offered to his person, especially the *groveling* one of erecting a gibbet before his prison-window, will, I expect, readily be believed, when I explicitly declare that I never heard of a single attempt to offer an insult, and that I had every reason to be convinced that he was treated by the officers around him, with all the tenderness and every civility in their power. I would fain ask Captain *Asgill* how he could reconcile such a belief (if his mind had been seriously impressed with it) to the continual indulgences and procrastinations he experienced? he will not, I presume, deny that he was admitted to his parole within 10 or 12 miles of the British lines; – if not to a formal parole, to a confidence yet more unlimited, by being permitted, for the benefit of his health and the recreation of his mind, to ride, not only about the cantonment, but into the surrounding country for several miles, with his friend and companion, Major Gordon, constantly attending him. – Would not these indulgences have pointed a military character to the fountain from which they flowed? did he conceive that discipline was so lax in the American army, as that *any* officer *in it* would, have granted those liberties to a person confined by the express order of the Commander in Chief, unless authorized to do so by the same authority? and to ascribe them to the interference of Count Rochambeau, is as void of foundation, as his other conjectures, for I do not recollect that a sentence ever passed between that General and myself, directly or indirectly, upon the subject.

  I was not without suspicions after the final liberation and return of Capt. Asgill to New York, that his mind had been improperly impressed; or that he was defective in politeness. The treatment he had met with, in my conception, merited an acknowledgment. None however was offered, and I never sought for the cause.

  This concise account of the treatment of Capt. *Asgill,* is given from a hasty recollection of the circumstances – If I had time, and it was essential, by unpacking my papers and referring to authentic files I might have been more pointed and full. It is in my power, at any time, to convince the *unbiased mind,* that my conduct, through the whole of this transaction was neither influenced by passion, guided by inhumanity, or under the control of any interference whatsoever. I essayed every thing to save the innocent, bring the guilty to punishment, and stop the farther perpetration of similar crimes: with what success the impartial world must, and certainly will, decide. –

With every great esteem and regard,

I have the honor to be, dear Sir,

Your most obedient servant,

G. WASHINGTON.

No. IV.

Copies of original letters and orders to American officers and others, respecting Capt. Asgill: extracted from General Washington's papers after the preceding letters were written.

To Brigadier Gen. HAZEN, Lancaster.

Head-Quarters, 3d May, 1782.

Sir,

The enemy persisting in that barbarous line of conduct they have pursued during the course of this war, have lately most inhumanly executed Capt. *John* [sic] *Huddy*, of the Jersey State troops, taken prisoner by them at a post on Tom's River, and in consequence I have written to the British Commander in Chief, that unless the perpetrators of that horrid deed were delivered up, I should be under the disagreeable necessity of retaliating, as the only means left to put a stop to such inhuman proceedings.

You will therefore immediately on the receipt of this, designate, by lot for the above purpose, a British Captain who is an unconditional prisoner, if such an one is in your possession, if not, a Lieutenant under the same circumstances from among the prisoners at any of the posts either in Pennsylvania or Maryland. So soon as you have fixed on the person, you will send him under a safe guard to Philadelphia, where the minister of war will order a proper guard to receive and conduct him to the place of his destination.

For your information, respecting the officers who are prisoners in our possession, I have ordered the Commissary of Prisoners to furnish you

Appendix Two

with a list of them. It will be forwarded with this. *I need not mention to you that every possible tenderness, that is consistent with the security of him, should be shown to the person whose unfortunate lot it may be to suffer.*

I am, &c,

G. WASHINGTON.

No. V.

To Col. ELIAS DAYTON, 2d. New-Jersey, Chatham.

Head-Quarters, 4th June, 1782.

Sir,

I am informed by the Secretary at War, that Capt. Asgill of the British Guards, an unfortunate officer, who is destined to be the unhappy victim to atone for the death of Captain Huddy, was arrived in Philadelphia, and would set off very soon for the Jersey line, the place assigned for his execution. He will probably arrive as soon as this will reach you: and will be attended by Captain Ludlow, his friend, whom he wishes to be permitted to go into New-York, with an address to Sir Guy Carlton in his behalf.

You will therefore give permission to Captain Ludlow to go by the way of Dobb's Ferry into New York, with such representation as Captain Asgill shall please to make to Sir Guy Carleton.

At the same time I would wish you to intimate to the gentlemen, that though I am deeply affected with the unhappy fate to which Captain Asgill is subjected, yet, that it will be to no purpose for them to make any representation to Sir Guy Carlton, which may serve to draw on a discussion of the present point of retaliation: that in the stage to which the matter has been suffered to run, all argumentation on the subject is entirely precluded on my part: that my resolutions has been founded on so mature deliberation, that they must remain unalterably fixed. You may also inform the Gentlemen, that while my duty calls me to make this decisive determination, humanity dictates a tear for the unfortunate offering, and inclines me to say that I most devoutly wish his life may be spared. This happy event may be attained, but it must be effected by the British

Commander in Chief: He knows the alternative which will accomplish it; and he knows that this alternative only can avert the dire extremity from the innocent, and that in this way alone the manes of the murdered Captain Huddy will be best appeased.

"In the meantime, while this is doing, I must beg that you will be pleased to treat Captain Asgill with every tender attention and politeness (consistent with his present situation) which his rank, fortune and connections, together with his unfortunate state, demand."

I am, &c.

G. WASHINGTON.

No. VI.

Extract of a letter To Col. ELIAS DAYTON,

2d New-Jersey, Chatham.

Head-Quarters, June 11th, 1782.

Sir,

You will inform me as early as possible, what is the present situation of Captain Asgill, the prisoner destined for retaliation, and what prospect he has of relief from his application to Sir Guy Carleton, which I have been informed he has made through his friend, Captain Ludlow. I have heard nothing yet from New-York, in consequence of his application.

His fate will be suspended till I can be informed of the decision of Sir Guy: but I am impatient, lest this should be unreasonably delayed. The enemy ought to have learnt before this that my resolutions are not to be trifled with.

I am, &c.

G. WASHINGTON.

No. VII.

P.S. To Col. Dayton

Appendix Two

Sir,

I am informed that Captain Asgill is at Chatham, without guard, and under no restraint. This, if true, is certainly wrong. *I wish to have the young gentleman treated with all the tenderness possible, consistent with his present situation.* But, until his fate is determined, he must be considered as a close prisoner, and be kept in the greatest security. I request therefore that he may be sent immediately to the Jersey line, where he is to be kept close prisoner, in perfect security till further orders.

No. VIII.

To Col. DAYTON, 2d. New-Jersey, Chatham.

Head Quarters, 22d June, 1782.

Sir,

I have received your two letters of the 17th and 18th instant. The only object I had in view in ordering Captain Asgill to be confined to the huts was the *perfect security* of the prisoner. This must be attended to. *But I am very willing, and indeed wish every indulgence to be granted him, that is not inconsistent with that.*

When I ordered on an officer for the purpose of retaliation I mentioned my willingness that he should make any application he thought proper to the British Commander in Chief, in whose power alone it lay to avert his destiny, but I, at the same time, desired it to be announced that I should receive no application nor answer any letter on the subject, which did not inform that satisfaction was made for the death of Captain Huddy. I imagine you was not informed of this circumstance, or you would have prevented Major Gordon's application on the subject.

I am &c.

G. WASHINGTON.

No. IX.

Postscript of a letter to Col. DAYTON, 2d New-Jersey, Morristown, dated

Head-Quarters, Newburgh, Aug. 25, 1782.

"P.S. You will leave Captain Asgill on parole at Morristown, until further orders."

No. X.

To his Excellency, General WASHINGTON, Commander in Chief.

Col. DAYTON's Quarters, Chatham,

May 17th, 1782. [432]

On the 30th of last month I had the honor of addressing your Excellency in writing, stating the manner of my confinement, and the circumstances that induced me to claim your protection. Being ignorant of the fate of my letter, it would be very satisfactory to me if your Excellency would be pleased to inform me if it has been received. In consequence of your orders Col. Dayton was desirous of removing me to camp, but being ill with a fever, I prevailed on him to let me remain at his quarters close confined, which indulgence I hope will not be disapproved of. *I cannot conclude this letter without expressing my gratitude to your Excellency for ordering Col. Dayton to favour me as much as my situation would admit of, and in justice to him, I must acknowledge the feeling and attentive manner in which those commands were executed.*

I have the honor to be,

with great respect,

your excellency's, most obedient servant,

CHARLES ASGILL. Lt. and

Capt. 1st Reg. Foot Guards.

Appendix Two

No. XI.

To Capt. LUDLOW, 1st Batt. British Guards, New-York.

Head-Quarters, August 5, 1782.

Sir,

Persuaded that your desire to visit Capt. Asgill at Chatham, is founded in motives of friendship and humanity only, I inclose you a passport for the gratification of it.

The inclosed letters for that gentleman, came to me from New-York, in the condition you will receive them: you will have an opportunity of presenting them with yourself. Your own letter came under cover to me *via* Ostend.

I have the honor to be,

Sir, your most obedient servant,

G. WASHINGTON.

PASSPORT.

Captain Ludlow, of the British Guards has my permission (with his servant) to pass the American post at Dobb's Ferry, and proceed to Chatham. He has liberty also to return to New-York the same way.

Given at Head-Quarters, the 5th of August, 1782.

G. WASHINGTON.

No. XII.

To Capt. CHARLES ASGILL, 1st Batt. British Guards, Prisoner, Chatham.

Head-Quarters, 7th October, 1782.

Sir,

I have to acknowledge your favour of the 27th of September.

The circumstances which produced in the first instance your unfortunate situation, having in some measure changed their ground, the whole matter has been laid before Congress for their directions. I am now waiting their decision.

I can assure you I shall be very happy, should circumstances enable me to announce to you your liberation from your disagreeable confinement.

I am, &c.

G. WASHINGTON.

No. XIII.

To Capt. CHARLES ASGILL.

Head-Quarters, 13th Nov. 1782.

Sir,

It affords me singular pleasure to have it in my power to transmit to you the inclosed copy of an act of Congress of the 7th instant, by which you are released from the disagreeable circumstances in which you have been so long.

Supposing you would wish to go into New-York as soon as possible, I also inclose a passport for that purpose.

Your letter of the 18th of October came regularly to my hands. I beg you to believe that my not answering it sooner did not proceed from inattention to you, or a want of feeling for your situation.

I daily expected a determination of your case, and I thought it better to await that, than to feed you with hopes that might in the end prove fruitless. You will attribute my detention of the inclosed letters, which have been in my hands a fortnight, to the same cause.

I cannot take leave of you sir, without assuring you, that in whatever light my agency in this unpleasing affair may be viewed, I was never influenced, through the whole of it by sanguinary motives; but by what I conceived to be a sense of my duty, which loudly called upon me to take measures, however disagreeable, to prevent a repetition of those enormities which have been the subject of discussion. And that this important end is likely to be answered without the effusion of the blood of an innocent person, is not a greater relief to you than it is to, Sir,

Appendix Two

your most obedient, humble servant,
GEORGE WASHINGTON.

APPENDIX THREE

Captain Charles Asgill's letter dated 20 December 1786, hidden for 233 years.[433] Italics by the author.

Capt Asgill^s Answer to
General Washington^s Letter &c
Address^d to the Editor of the
Newhaven Gazette

London Decr 20th 1786

Sir /

In your Paper of the 24th August [actually 16 November] the publication of some letters to & from Genl Washington together with parts of the Correspondence which pass^d during my Confinement in the Jerseys renders it necessary that I should make a few remarks on the insinuations contain^d in Genl Washington^s Letter, & give a fair account of the Treatment I received while I remain^d under the Singular circumstances in which Mr Washingtons judgment & feelings thought it justifiable & necessary to place — the extreme regret with which I find myself oblg^d to call the attention of the publick to a subject which so peculiary if not exclusively concerns my own Character & private feelings will induce me to confine what I have to say within as narrow a Compass as possible —

very little is necessary to the *anonymous* letter of the American Correspondent, who boasts his introduction to *Coffee house Sages* & making his assertions on *coffee house authority* so confidently affirms that Charges were exhibited against General Washington, by Young Asgill of illiberal treatment and cruelty towards himself that he alledges the delay & the release to Count *Rochambeau* This affair (he adds) is stated by Asgill & canvass^d at the British Court as before related — *Vide No.2

Appendix Three

It is sufficient to say that this Gentleman whoever he is never took the pains of ascertaining the truth of the intelligence he received from his Coffee house sages, by an *application to me*, tho I almost resided Constantly in London & that by neglecting so to do he has expos^d himself to the degrading circumstance of having positively asserted a *groundless falsehood* for I never did either suggest or countenance the report of a Gibbet having been erected before my prison Window — a Prison was indeed a Comfort that *was denied me* nor had the fact been so would it among the many indignities & unnecessary hardships I endur^d been a source of uneasiness at the time or a subsequent object for Complaint weak indeed must that mind be which in a situation similar to mine could feel additional distress or a more strong impression of what was to be expected by the exhibition of such an object — in Truth no Gibbet was erected in sight of my window Tho during my Confinement I was inform^d at different times & by various persons that in Monmouth County a Gallows was Erected with this inscription "up goes Asgill for *Huddy" for the truth of this I cannot vouch as I never saw it myself the report of it at the time gave me no anxiety I disclaim it therefore as a subject of Complaint & willingly consider it if it existed as one of the necessary means of *keeping the American Army in Temper —*

*the American Lieu^t who suffered Death within the British Lines

I never ascrib^d what the American Correspondent call^d the Sentence to the interference of Count Rochambeau still less did it ever occurr to me to attribute it to the *humane procrastination of General Washington*, had I been inclined to do so his letter of the 11th of June to Col^l Dayton would have been sufficient to dispell that error as I should then have found him declaring "that my *fate* should be *suspended* till he heard from Sr Guy Carleton, but that he was *impatient* least this should be unreasonably delayed"

*Vide No.6 extract of a letter to Col Dayton

I have ever attributed the *delay* of my execution to the *humane, considerate & judicious conduct* of Sr Guy Carleton, who amus^d Genl Washington with hopes & sooth^d him with the Idea that he might obtain the more immediate object of retaliation &

The Charles Asgill Affair

Vengeance this Conduct of Sr Guy produced the procrastination which enabled the French Court particularly Her Majesty [Queen Marie Antoinette] to exercise the characteristic humanity of that great & polish^d nation in its most exalted & beneficent light uninfluenc^d by policy or Annexion, not check^d by national Hostility, or retard^d by the insignificance of an individual the great object of military Glory, the ease & prosperity of a people the protection of *allies* & the judicious pursuit of public advantage have been generaly the end of the attainment of Kings & their Ministers but it was reserve^d for the present Sovereign of the French Nation to look with Compassion from the heighth of Empire, on the unjust oppression of a single Victim taken from among hostile troops & devoted to the blind Vengeance of their vindictive allies — it was their benevolent interposition that thwarted the rage of Sanguinary resentment, stop^d the uplifted & impatient hand of retaliation, to them, seconded by the humane & tender sollicitude of a Minister, whose heart did honor to Mankind & whose abilitys were a pride & bulwark to his nation I owe my release from a Death to which the faith of Nations & the sacred Articles of a Solemn Convention forbad my being expos^d this would have appear^d in the letters which pass^d relative to my Confinement had Mr Humphreys (who declares himself acquainted with the minutest Circumstance relative to that *unpleasant* affair) *thought proper to publish the whole of that Correspondence** — the 14th Article of the Capitulation made at York Town expressly says "that no Article of this Capitulation shall be infring^d on pretext of Reprisal" Mr Washington felt so strongly the force of this provision that in his letter to Genl Hazen the 3^d May 1782 "he expressly desires that a British Captain who is an *unconditional* Prisoner (if such a one is in his possession) if not a Lieutenant under the same Circumstances may be designated by lot as an object of Reprisal" a Lieutenant is here to be taken for the purpose of reprisal rather than not take an unconditional Prisoner, if it was not an act of injustice to take a Captain under articles of Capitulation why was a Lieutenant to be taken as a reprisal for the loss of a Captain [If] it was in isolation of public faith & an act of Tyranny could the want of a proper object [line crossed out] of reprisal justify such an exercise of Power for the sanguinary gratification of eager Vengeance * Comte de Vergennes   Indeed Mr Washington^s suppression of a letter

Appendix Three

subsequent to that [just] quoted in which he demands a Brittish Captain taken under Articles of Capitulation shows how unwilling he is to fix the attention of the public (to whom he now addresses himself) in that part of his Conduct — he felt from the first how impossible it was to justify this act of Power by any arguments grounded on the Law of Nations & in his letter of the 4th June to Col^l Drayton he declines entering into any discussion of the present point of retaliation I however thought it my duty in justice to myself & the Army in which I serv^d, to write* the following remonstrance to Mr Washington when I was first denoted as the victim of his Resentment. *vide

I will go no further into the discussion of the injustice of my Confinement but will now enter upon the nature of it & the treatment I receiv^d while it lasted. The ready belief, which Mr Washington gives to the [Report] of M^r [Tighlman]'s credulous Correspondent needs no further comment I have already observ^d upon that Circumstance, & the forfeiture of Mr Washingtons good opinion would give me greater uneasiness than I am at present despos^d to feel from it, had the possession of it been attended with more favorable treatment whilst I remain^d in his power.

M^r W has ask^d me two questions in his letter to the first (relative to the Gibbet) I have already given an answer to the second relative to the permission I receiv^d to [ride] — I reply that I did indeed receive such a permission *after four months close confinement* when my Health had been impaired & my Constitution reduc^d by *a long series of Hardships & insults* —

I will now beg leave to ask a few Questions of General Washington to which I also presume he will not give any negative — was I not publickly exhibited & unnecessarily paraded with an Escort of Dragoons in to Philadelphia & the Towns in my way to the Jerseys & was I not confin^d in a small room of a public House to which all persons were admitted to View me who paid for the gratification of their Curiosity by calling for Liquor [for] the sentinels of the door Tho they had orders to watch me narrowly, had none to prevent the *insults I daily experienced* does the General know that there have been frequently numbers of drunken people surrounding my bed during the night was I not refus^d to be confin^d in a private house in the same village tho [many] of the inhabitants would readily have receiv^d me & that I might have been guarded with equal security & was not the reason assign^d for this refusal that *I should cease to be a spectacle to the people if I was privately confin^d* & did I not in consequence of my irksome situation write the following letter [?] did not general Washington detain the letters sent by my Family tho they contain^d nothing political & were merely calculated to relieve my mind from the anxiety & distress under which I labour^d, from the dreadfull accounts which were spread of various calamitys afflicting my family in consequence of the situation in which I then was & *were more than two ever sent to me & were they not sent a considerable time after Genl W first receiv^d them*

did not Genl W delay acquainting me with the Act of Congress, for my Release, for five days after it pass^d & was I not at last sent to Dobbs Ferry without an Escort expos^d to the insults of the people who wish^d to have seen me sacrific^d to the Memory of Capt Huddy. These Circumstances G W must have Known, he may perhaps have also heard that the provisions delivered to me at the early period of my confinement were exceedingly bad till the Landlord procur^d me better nourishment at an exorbitant price upon finding that I had a considerable sum of Money. I was repeatedly & almost continuously *vide insulted not indeed by the Officers who guarded me They & *particularly Col Drayton behav^d civily & I was beat violently & cruelly* for refusing to answer the impertinent questions put to me by a visitor who thought himself entitled to be satisfied on every particular having paid for his admission — These were the Comforts I enjoy^d These were the attentions I receiv^d from General Washington —

I had however a comfort beyond his reach — totally above his power to invade it was the Pride, the Consolation & Support which I deriv^d from *the exalted Friendship & Kind Compassion of Major Gordon of the 80th Reg^t* who feeling for the distresses of a Brother Officer, [entrusted] by the best of Hearts in the cause of humanity & unwilling to leave a Youth of eighteen unadvis^d & unsupported to act in so peculiar & difficult a situation, sacrific^d every Comfort to partake my hardships & Confinement & by the impulse of his excellent & noble Heart, felt on the [first] acquaintance all the steady & persevering Zeal that the longest & most tried Friendship could hope for or Claim he was there the whole of this transaction the partaker of my hardships the support

Appendix Three

of my Spirits, & the monitor of my conduct *I am delighted at having the opportunity of proclaiming to the World his generous & benevolent attentions* [233 years passed before these words reached the World] Tho whilst I do justice to his Memory I aggravate the sensations of regret, which I must ever retain, for the loss of him

I have now only to answer that part of G W^s letter in which he expresses his opinion of the acknowledgements which were due, for the treatment I had receiv^d & his suspicions on my silence — *I leave for the public to decide how far the treatment I have related deserv^d acknowledgements — the motives of my silence were shortly these The state of my mind at the time of my release was such that my judgment told me I could not with sincerity return thanks my feelings would not allow me to give vent to reproaches*

A note in the *Journal of Lancaster County's Historical Society* reads:

> Asgill included nine footnotes in his letter. Two were complete, and three referenced the 1786 newspaper article. The other four may have referenced enclosures, but we have not been able to find information about them.[434]

The Asgill letter was posted from London to New Haven, Connecticut, in December 1786, and it had surfaced for sale in New Haven in 2003. I jumped to the (wrong) conclusion that it had never left there. So I got in touch with Nick Aretakis, of William Reese Company, antiquarian bookseller, New Haven, who had acquired it in 2003, and sold it in 2008, saying:

> If I may refer you back to correspondence we had in 2009, following your sale of the above item for $16,500, I am wondering whether you would be so kind as to tell me the provenance of this item? I am curious to know how and why it surfaced after more than two centuries, and would just love to know the journey travelled by those documents.

What Mr. Aretakis told me stunned me! He wrote:

> We bought the volume nearly 20 years ago from a bookseller in the United Kingdom (I do not recall who) and, as I

mentioned to you in 2009, sold it to a private collector. As that collector has since passed away and the library has been dispersed at auction, I can tell you that it was sold to the collectors Ira and Barbara Lipman of New York City.

He suggested I contact Selby Kiffer, International Senior Specialist, Books & Manuscripts, Sotheby's, New York. I emailed him, and was told that Mr. Lipman had passed away in 2019, and his wife, Barbara, in early 2020. Their possessions were sold at auction on 13-14 April 2021, realising US$16 million. However, no bids were received for the letterbook,[435] and it was, therefore, being retained by the Lipman heirs.

Ira Lipman was an American businessman and philanthropist. He was the founder and chairman of Guardsmark, a privately owned security company. He died on 16 September 2019, aged 78, in Manhattan, New York City.[436]

I was extremely disappointed to discover that Asgill's attachments, and additional comments at the end of his 1786 letter are no longer with the letterbook, but what they had been can be speculated; he had written:

> I however thought it my duty in justice to myself & the Army in which I serv^d, to write the following remonstrance to Mr Washington when I was *first denoted* as the victim of his Resentment. *vide* the end of the narrative

This can only mean the letter Asgill wrote to George Washington from Philadelphia on 30 May 1782,[437] three days after lots were drawn, in which he "remonstrated", as follows:

> Perfectly innocent of Captain Huddy's Death, & even to this moment uninformed of the circumstances & ever having acted consistently with the Tenor of my Parole I am certain in Justice his Death can never effect me, nor do I know why my Life should be an Atonement for the Misdemeanours of Others—I claim protection under the 4th Article of the Capitulation & from your Excellencys known Character I have every Right & Reason to expect it.

He wrote "4th" Article erroneously – it was the 14th Article. However politely Asgill had expressed himself, it was certainly also a remonstrance, so this he must have copied out at "the end of the narrative."

Appendix Three

In the two letters Washington wrote to Asgill he never acknowledged the "Horrors" and "sufferings" Asgill had brought to his attention, nor is it known if he ever addressed the source of these conditions Asgill had explained. Perhaps it was deemed that the people of Monmouth County were owed the opportunity to vent their feelings on Asgill? But on 13 November 1782, Washington wrote, releasing Asgill from captivity, saying: "Your letter of the 18th of October came regularly to my hands. I beg you to believe that my not answering it sooner did not proceed from inattention to you, or a want of feeling for your situation."[438] So, he had known, all along, but did nothing to spare Asgill his six months of deplorable treatment at the hands of the tavern's clientele. In that letter of 18 October, Asgill *had also remonstrated* with Washington, saying:[439]

> I hope your Excellency will give orders that my Parole may be withdrawn, as that Indulgence without a prospect of further Enlargment affords me not the least satisfaction, I had rather indure the most severe confinement, than suffer my Friends to remain as at present deceived, fancying ever since my first admission on Parole, that I was entirely liberated & no longer the Object of retaliation — if your Excell^y could form an Idea of my sufferings I am convinced the trouble I give would be excused.

So, this "sufferings" letter was undoubtedly also one of the now missing attachments Asgill included in his 1786 letter. These mentioned (and very likely his letter to Washington of 27 September, speaking of the "Horrors") were the documents he wished to put forward in his own defence. Sotheby's sent me a photograph of the first and last page of Asgill's letter, which tallied perfectly with what I had already obtained, so this is confirmation that they have disappeared.

As to how this letterbook arrived back in England, speculation and conjecture are our only tools, and my thoughts on the matter are:

Asgill wrote to the *New Haven Gazette* on 20 December 1786 and despatched it (along with his attachments) to the Editor, who failed to publish it. Josiah Meigs bound it with the Washington papers, already in his possession. Perhaps, sometime later, Asgill wanted it back (once he realised it would never be published) and requested a friend to collect it from New Haven, bringing it back to him on his next visit to England – or even posting it back to him. It then wound its way down through the centuries, to relatives, and descendants of relatives (most likely the Ogles) to eventually be given to a British bookseller in 1943. The last Ogle

The Charles Asgill Affair

Baronet, Sir Edmund Asgill Ogle, 8th Baronet (1857–1940) died without issue in 1940 – during World War II. In 1943 Maggs Bros Ltd., 48 Bedford Square, London, (one of the longest-established antiquarian booksellers in the world, established in 1853 by Uriah Maggs),[440] acquired it and listed it for sale that year.[441] It was priced at £21. In their 2021 auction, Sotheby's had estimated its value at between $15,000-$20,000.

Or perhaps it had been sold, in New Haven, decades earlier than 2003, and had been purchased by a buyer in England? That would be a simpler solution. But I found no "for sale" records going further back than 1943.

Did Maggs sell it prior to 2003, or was it in their custody, unsold, for 60 years? They tell me that they have no records which would confirm or deny this, so, to my knowledge, no trace of the letterbook's movements emerged again until 2008, when William Reese sold it to Ira Lipman – and again in 2021 when his heirs listed it for auction by Sotheby's. My 2022 "close encounter of the potentially expensive kind" has only added to the enduring mystery of what happened to Asgill's letter, and why it was never published until 2019, in Lancaster, Pennsylvania. Sotheby's told me that it is normal for the provenance of letterbooks to have their movements recorded at the back of the book; but this letterbook has no record of any kind. Since the letterbook had failed to sell at Sothebys, I hoped that perhaps an after-auction offer might be successful. I suggested $3,000. This was not accepted.

Imagine my absolute astonishment when, later in 2022 I discovered that the Lipman heirs had put the letterbook up for auction *again*, eighteen months later, with Doyle, of New York, to be auctioned on 16 November 2022, estimated at between US$8,000 and US$12,000.[442] I was in shock to be confronted, for the third time, with the potential to acquire it. Through my own misunderstanding of the listing, I erroneously thought that the reserve price was $4,000. I got in touch with Doyle, discovered just how much the extra costs would be, contacted a shipping agent in New York City to establish what sending it to me would entail, and put in my bid of $4,000 (aware that the final cost to me would be in the region of £5,200, but wasn't entirely sure what I would have to pay for import duty in the UK). I registered for online bidding, in order to watch the auction live, on 16 November. It was all over in seconds and I was once more shocked to discover that there were no bids of any kind, other than mine. My heart raced, my imagination ran riot, and I visualised myself seeing and touching the most important letter Asgill ever wrote. I

Appendix Three

was certain that, having bid what I believed was the reserve price, that it would be mine. How could it not be? It was my understanding that the heirs would have to accept my bid. I was wrong on all counts.

I may not live long enough to know what happens next, although I did email one of the heirs to put in another offer. No response was received.

APPENDIX FOUR

An eye-witness account of the drawing of lots on 27 May 1782

Letter of Captain The Hon. Henry Greville, of the 2nd Foot Guards [mistranscriptions corrected], to his Mother, transcribed, by courtesy of the [7th] Earl Spencer, from the Archives of Althorp, Northampton[443]

Lancaster, Pensylvannia,

May 29th, 1782.

My Dear Mother,

probably before this reaches you, you will have heard of the strangest Transaction that ever happened between two Civilized Nations. As a party concerned I will however state the true Acc^t to you, a Captn Huddy was hung sometime since by our Refugees and a label was fixed to his breast signifiying they retaliated upon him for the Death of a Captain White a Loyalist who they pretend had been unjustly put to death by the Enemy, not contented with hanging this unfortunate Man within their own lines. nothing would satisfy them but they must carry him over at night to the Jersey shore and leave his Body there; this exasperated the Inhabitants very much, and Complaints were made to Genl Washington who immediately demanded of Sir H. Clinton the perpetrators of the above inhuman act. I saw Genl Washington's letter, it is very strongly worded and in it he says to save the Innocent I demand the guilty. Sir H. Clinton in answer finds fault with Genl Washington's improper expressions, as he terms them, and tho' he assures him he had ordered a Court of inquiry to sit upon the people who had been guilty of this Action, which he says was entirely without his Knowledge, yet he does not say whether he will give them up or not, in this state were Matters when a letter came to Genl Hazen, Commanding at this post, to summon all the British Captains at York and Lancaster and make them draw lots who should be given up as he meant to retaliate on one of us, being the only means in his power to put an end to our barbarous method of proceeding, never was surprise equal to ours, little did we think an Article of so Solemn a Capitulation as ours was, concluded

Appendix Four

between three Powers, would be so shamefully violated, however there was no Alternative so we all set off from York Town to Lancaster, which was 20 miles off, we were but thirteen Cap[ms] in the whole and five of those belonged to the Guards, on our arrival we all repaired to Major Gordon's House, he is our Commanding Officer here, we there learnt for a Certainty that one of us must be given up and that the next morning was the time appointed for us to draw lots; we were all unanimous, as you may suppose in refusing to draw ourselves, we said we had been guilty of no crime and deserved no punishment, we were in their Pow'r, they might do what they pleased with us, but that we positively would not lend a helping hand to our own destruction, this was the Evening of the 26th. I can't say I slept remarkably well that night, on the Morrow we were all summon'd to the Major's House, he there informed us that at ten o'Clock all the British Captains were ordered to repair to the Bear Inn, where General Hazen would meet them. The momentous hour arrived and we repaired to our Guild Hall, we were ushered into a little Room with a great deal of Ceremony and Genl Hazen, attended by his *Aid-de-Camp*, Adjutant and an officer of Dragoons, joined us almost instantly; every tongue was silent on his Appearance, and he, with Appologies for his Errand, informed us that it was General Washington's order that one Captain should be given up, and that he was to be pitched upon by Lot, we refused to draw – he replied then the lots must be drawn for you – Major Gordon stood up and begged his patience for a few moments – with infinite feeling he expressed his Sorrow and Amazement at so unexpected a transaction taking place, he urged the Injustice of the Innocent suffering for the Guilty. Whatever action, said he, this barbarous and Inhuman method of carrying on the War regains, surely it does not extend to us who, by the 14th Article are expressly under your protection, however, said he, since you must Obey your Orders, I have to request a passport for an Officer to go immediately to Philadelphia with such letters as I shall chuse to write, and I likewise hope he may be allowed to proceed to New York with letters for Sir Guy Carleton. General Hazen whose behaviour throughout the whole affair has been Noble and Generous, immediately granted his request and again appologized for being Obliged to execute his orders, he said he was but a Servant and as such must behave, he then Ordered the Lots to be made, and

The Charles Asgill Affair

desired we all would be present at the drawing them. Altho' I have felt many more disagreable sensations upon other occasions than I did upon this, yet I can assure you my mind was in a very uneasy state for above half an hour while they were calling out the Lots, during which time we sat in a Circle, where there was almost a dead silence observed – Judge yourself what all our feelings must have been at that time, I cannot think the most bloody Field of Battle could wear half the terrors this hour did – to keep you no longer in suspense, two hats were brought in, in the one was all our names, in the other twelve blanks and one piece of paper with *Unfortunate* wrote upon it. eight lots were drawn and as many blanks, there remained our five to draw, I think but five, I'm not quite certain, the ninth my name was mentioned and no poor Criminal ever waited the verdict of a Jury with more impatience that I did for mine. It was a Blank, and tho' to outward appearance I was not much affected by it, yet I never was much happier in my life, and in silence I returned my thanks to that God who protected me – no longer allarmed for my own safety I began to be anxious about the fate of my Friends, two particular ones Captains Ludlow and Asgyle of the guards who were still to draw, and it was with inexpressible concern that I saw the Unfortunate lot fall upon Asgylle, he is an amiable young man perfectly adored by his Parents, esteemed and respected by his friends. When the Commissary of prisoners mentioned the name of the Sufferer he burst into tears, and for some time could not articulate, we were all indeed very much agitated, Major Gordon in particular, never did a man possess more exquisite sensibility than he does; he is an honour to Society, and his attention and care respecting everything that could alleviate my poor Friend's misfortune will ever be remembered with Gratitude and esteem by his Brother Officers. Asgylle, tho' evidently affected with his Fate, yet bore it with uncommon resolution, he thanked Gen[l] Hazen in the prettyest style for the politeness he had shown him, and took his leave. I walked home with him and endeavoured to keep up his spirits, at first he was melancholy and thoughtfull, seemingly very unhappy, but it gradually wore off, and no person to have been in his company could have supposed he was doomed to die. Ludlow was dispatched immediately to Congress and to the French Minister with letters, and Gen[l] Hazen wrote the prittyest one imaginable to Gen[l] Lincoln in favour of Asgylle and altho' it was the Gen[ls] Order

Appendix Four

to send off the Officer on whom the Lot fell immediately to Philadelphia, he per-mitted him to stay all the 27th and till seven O'Clock the next morning in Order to give time to Ludlow to present his letters, and then did not send a Guard with him but rode out of Town himself with Major Gordon and Asgylle, for privious to this the Major had got leave to go with Asgylle to keep him Company, subjecting himself to the same confinement. He had not been gone ten minutes before I rec^d the enclos'd and this token of his Gratitude will remain fixed in indelible Characters on my mind and he has proved that his exquisite feelings are equal to his Magnanimous fortitude. We are all at this moment waiting with anxiety to know his fate, we think and hope the confinement and anxiety of mind will be his greatest punishment, every Person in this Town was affected at his Missfortune. There were more tears shed here the 27th May than ever fell on any occasion. I will not close this till I hear everything is settled

July the 13th

Tho' it is currently believed that Mr. Lippincote is to be delivered up, yet we know nothing for certain, except that our Unfortunate Friend is still living. I'm obliged to send my letter off now as I might not have another Opportunity this Age

Copy of Cap^{tn} Asgylle's letter to Cap^{tn} the Hon. Henry Greville of the Second Foot Guards, enclosed in Captain Greville's letter of May 29th [mistranscriptions corrected]

Lancaster, May 28th, 1782.

Dear Greville,

I should be unpardonably remiss did I not leave behind some token of gratitude for the feeling Attention, the inexpressible grief which your manner and Countenance in defiance of all efforts to conceal did undesignedly betray – you seemed to participate my Missfortune and from your Soul to pity my Situation. I only wish at that time I had possessed sufficient power of Utterance to have Alleviated your distress and to have acknowledged my Sensibility of your conduct – this my Dear

Greville is the only method I can take of telling you how much I am and ever shall be

Your Obliged and Faithful Friend, Charles Asgylle.

APPENDIX FIVE

Lady Asgill's letter to the comte de Vergennes

To the comte de Vergennes, King Louis XVI's Minister of Foreign Affairs, from Lady Asgill, 18 July 1782

If the politeness of the French Court will permit an Application of a Stranger, there can be no Doubt but one, in which all the tender feelings of an individual can be interested, will meet with a favorable reception from a Nobleman whose character does honor not only to his own Country but to human Nature. The Subject, Sir, on which I presume to implore your Assistance, is too heart-piercing for me to dwell on, and common fame has most probably informed you of it; it therefore renders the painful task unnecessary.

My son (an only Son) and dear as he is brave, amiable as deserving to be so, only nineteen [twenty], a prisoner under Articles of Capitulation of York-Town, is now confined in America an object of retaliation! Shall an innocent suffer for the guilty? Represent to yourself, Sir, the situation of a family under these circumstances: surrounded as I am by objects of distress; distracted with fear & grief; no words can express my feelings or paint the scene - my husband given over by his physicians a few hours before the news arrived, and not in a state to be informed of the misfortune - my daughter seized with a fever and delirium, raving about her brother, and without one interval of reason, save to hear the heart-alleviating circumstance let your feelings, Sir, suggest and plead for my inexpressible misery — a word from you like a voice from Heaven will save us from distraction and wretchedness. I am well informed General Washington reveres your character; say but to him that you wish my Son to be released and he will restore him to his distracted family, and render him to happiness. My Son's virtues & bravery will justify the deed. His honour, Sir, carried him to America - he was born to affluence, independence and the happiest prospects — let me again supplicate your goodness, let me respectfully implore your high influence in behalf of innocence — in the cause of justice — of humanity — that you would dispatch a letter to General Washington from France and favour me with a copy of it to be sent from hence. I am sensible of the liberty I take in making this request, but I am also sensible that whether you comply with it or not, you will pity the distress that suggests it Your humanity will drop a tear on the fault and efface it. I will pray that Heaven may grant you may never want the comfort it is in your power to bestow on

(Signed) Thérèse Asgill[444]

APPENDIX SIX

Jean-Louis Le Barbier's Play

During the Asgill's 1783-4 visit to Paris, Lady Asgill was presented with a manuscript play titled: *Asgill, Drame, en Cinq Actes, en Prose*. Written by Jean-Louis le Barbier, it was not published until 1785. The price of the London edition was £1.10 shillings.[445] The cast of characters included:

> Asgill English Officer
>
> Madame Asgill, Mother Of Asgill
>
> Colonel Gordon, Friend Of Asgill
>
> Washington, American General
>
> An Officer, French General
>
> A Deputy Member Of The American Congress
>
> Laurence, Gordon's Valet
>
> Henry, Madam Asgill's Valet
>
> Nancy, Madam Asgill's Servant
>
> Tranche-Montagne, French Grenadier
>
> An American Grenadier
>
> A Virginian Farmer
>
> An Old Virginian Man
>
> A Virginian Woman, Daughter-In-Law Of The Old Virginian Man
>
> A Child, Son Of This Woman, Grand Son Of The Old Man
>
> A Junior Officer, American
>
> American And French Soldiers

In the play, le Barbier le-Jeune shows Washington plagued by the cruel need for reprisal that his duty requires. Washington even takes Asgill in his arms and they embrace with enthusiasm. In his introduction, the playwright explained that he had hoped to perform the play in the family's honour, and in their presence, but had been uncertain when they would be leaving Paris. His sentiments come through clearly in his introduction, in spite of what follows being a translation from the eighteenth century French and coming, as it does, un-paragraphed. Le Barbier wrote:

> Obstacles that I was able to foresee, and motives that I needed to respect, having prevented the launching of this Play in 1783, I had decided to put it to one side, until I was convinced by public interest that the subject should be presented in the French Theatre. I do not want to allow myself any details on the way which gave its author the gift of knowledge with which I had already treated this subject; I want even less to argue the advantage of its success. What I have already obtained in this manner, must necessarily have inspired him with the confidence to believe that approbations will meet in his support; I left them to it; it is not difficult to do much better than me; I am not a man of Letters; faults which appear in my Play, are proof of this. I do not dissimulate myself from this, it is why I dare to hope for some indulgence from all good people, that taste and enlightenment stemming from the quality of a good heart, will not judge me harshly. In dealing with this interesting subject, I have satisfied my sensibilities; it is this that has inspired me from the first time I picked up a pen; it is this which, making my heart beat through all the agitation that caused the heartrending idea of seeing the wretched virtue, saw me devote sleepless hours to this work and to the detriment of my other interests. These moments, time and again, difficult and gentle, never faded from my memory; they come back to me repeatedly, the greatest joy for a sensitive spirit, they bring tenderness to the ills of his fellow men. There are, I say it again, serious faults with this Work; but let one feel able to forgive me these in favour of the thought that this has inspired me; I would have liked to have minimised them before publishing, but my state of mind and my journalistic studies no longer permit me to share my time. This that I claim as a just title in the actual moment, this was to have been the first to put on Stage a subject so touching; this is the only advantage to which I pretend, and that I will not give up to anyone, because I can supply the most convincing evidence, that my Play has been performed before the best Company in Paris, for more than eighteen months. I say to you here with pleasure, that in the time that I have been involved,

Appendix Six

I am indebted to judicious comments from Messieurs de Courcelles and Grangé (Actors from the Italian Theatre), as distinguished for their talents as their knowledge, who have ignored a great many faults: it is necessary to warn my Readers, that outside the plight of the young Asgill, the *Mercure de France* was the only period Work that time allowed me to read; it is to this alone that I worked on my subject; for me it never became one, in this painful situation, that led me to have serious misgivings; that of finding there the account of the tragic end of this fascinating victim, often made me change the script; suffice it to say finally, the moment when I focused on his happy release, gave me cause for pure joy, that in the soft emotions where my heart was beating, I did not stop saying to myself; that this story represented a great subject for the Theatre! [let me, your author, add, this subject would make a great film!] The means of displaying to the World the charitable virtues of our beloved Royal Family. Everything that I had learned of interest about the Americans, their enthusiasm, their up-front earnestness, from the roots of their Country to the presence of the French army, displayed all that compassionate comradeship, recognition inspired in these Unfortunate People for Friends and generous Liberators; all this, I say, repeatedly added new interest which inflamed my imagination to the point where I dared to undertake it. The difficult work that it has caused me since, has made me realise that emotions stemming from sensitivity do not always hold sway over talent. Truth and modesty prevent me taking for myself until a certain place where M. de Mayer says in the PS to his Letter to the Editor of his laudable Novel, "The arrival of Asgill (this is his expression) has inspired two well-known and distinguished writers, the decision to draw on part of my Novel, etc etc". And further on, he finishes by saying: "these two Writers have certainly attracted the attention of the Public, and I am flattered that they have drawn on such an interesting subject". It is likely that this is the case, but I exclude myself from this number because it would be ridiculous for me to lay claim to this Writer's qualities. One could not claim this title without impunity. To merit it, one would have to be dignified by being recognised for distinguished accomplishments; and, in other areas of Literature, there is nothing one could do to prove what had appeared under my name. If I owed the introduction of my Play to the interesting Novel by

M. de Mayer, justice and decency would want me to express my sincerest thanks; he obtained this without doubt from the two Writers that he cites, when they published their Works; as for me, I will confine myself to assuring him, along with my Readers, that, when Madame Asgill arrived in Paris, it was more than six months since my Play had been finished, and that, for want of knowing when she would be leaving, I was unable to have the honour of presenting it to her, as I had hoped; it is not, then, her arrival which inspired me to write this Play; the Letters that I have been honoured to have been sent by her, and the Dedication of it that I made to her immediately on her return to London, and what one will discover at the conclusion of the Play, are again so many more pointers in evidence of what I am saying.

After reading the play, Lady Asgill felt there were a few small matters the writer would appreciate knowing, and in particular she wanted the fictitious colonel of the play to be renamed as Colonel James Gordon (Le Barbier made this amendment). She also felt her servant should be more appropriately named – Nancy was suggested and again Le Barbier complied with her wishes. These comments were conveyed in the following letter:

Letter From M. Brilly [Elie Brilly, the French Huguenot Pastor of St. Jean (French Church), John Street, Spitalfields and close confidante of Sarah, Lady Asgill]

To the Author.

London, 30 January 1784.

Sir, My Lady Asgill, who has honoured me with her confidence and friendship, asked me to acknowledge the receipt of your Play by Lady Clarges [Louisa, Lady Clarges (1760–1809) was an accomplished singer and harpist]; I could not acquit myself more voluntarily of this commission, because she has provided me with the opportunity to make the acquaintance of a person of distinguished attainments, and whom I would wish to know better. The subject, interesting by itself, will not be able to fail to grow more, having been developed by an imagination as fertile as yours; and to paint feeling so well, it is needed, Sir, to be possessed moreover of a heart above common mediocrity. I am not trying to make eulogies here, your modesty stops me; but I wish to speak here of facts. The Play has been seen, and actually circulated

Appendix Six

amongst the best companies; the whole world is enchanted. The Asgill Family, one of the most loved in this country, dedicates to you always its esteem and gratitude; and it awaits happily the moment when it will be able to prove this in a less ambiguous way, and which will be most agreeable to you. Deign, Sir, to give me a confidence in this regard, I will anticipate myself (in case you are determined to publish your excellent Play) on certain necessary changes that are relevant to the Actors. It is not M. Murray (*), but Colonel Gordon, (since deceased) who was the Confidant & Mentor for the unfortunate Asgill. You gave to the Confidante of Madame his mother the name of Folly; the word is synonymous with that of folly in French; you wanted to say without doubt Polly, which is common to waiting-women. As she sustains a role more elevated, you are able to call her Clarissa or Nancy. Allow me to point out again that it is not Sir Guy Carleton, but Sir Henry Clinton, who commanded at that time in America; the first was Governor of Quebec.[446]

These remarks, which I take the liberty to make, do not diminish in any way the intrinsic merit of your production; and following the maxim; that it is easier to criticise, than to follow at a distance those that are criticised; I would hazard, if you would permit me, on another occasion, some further little observations.

It remains to me to assure you, that the very lovely Asgill Family, who implored me to ask you to accept the assurances of its gratitude and of its most perfect esteem, to be able to support your views and make the Play public, and to prevent others from profiting from your ingenious labours, you will be able to reclaim for yourself from it wherever, and whenever.

I have the honour to be, with the sentiments that have inspired me with your generous enthusiasm for suffering people, and your superior talents in portraying these misfortunes.

Sir,

Your very humble and very obedient Servant,

Elie Brilly.

Le Barbier added the asterisk shown in the letter above, and wrote:

The Charles Asgill Affair

(*) The name of Murray, for which I have substituted (after this notification) that of Gordon was absolutely an ideal person, whom I made up to establish and facilitate the intrigue of my piece; because I was ignorant of the fact that young Asgill had in reality, in the person of Colonel Gordon, a generous defender. I had supposed that this character could be in the range of possibilities, and who effectively existed. It is very correct to place here the real name, and to render public homage to the sentiment which animated Colonel Gordon's heart for his unhappy friend.

Le Barbier apparently responded to Lady Asgill directly, for he recorded her subsequent response:

Madame Asgill's Letter to the Author.

London, 31 July 1784

Sir,

I was surprised, no less, to have the honour of receiving your letter; I was ill in bed, when your two first ones were delivered with the Play, Workmanship which brought back so strongly unhappy sensitivities to which I had been subjected, not to open the wound that a skilful and generous hand had already so happily closed, and so judged by those to whom I ill advisedly gave it to read, in the circumstances in which I found myself, my son at the time being away from the Capital, but even less familiar than me with the French language. I asked Mr Brilly (a Clergyman) to tell you how much I appreciated the honour that you have been well pleased to do me, and to assure you that I anticipate with pleasure the means that will effect my recovery when I can show you my gratitude, and will not fail to do this as soon as possible. M****, who went to Italy, was commissioned to hand you my Letter(*) in person; this is what persuaded me that you must have received it some time ago. It is very likely that he will visit you yet, and that some unexpected accident on the way has been the cause of his delay. Had I the advantage of being familiar with you, Sir, I am persuaded that you could well wish to render me the justice of believing me incapable of such a fault towards a person of your attainments, and to whom I am so obligated. If I was guilty of such negligence, the memory alone would be a more severe punishment for me than all that you would be able to wish for me, and you would excuse me sooner than I

Appendix Six

would have forgiven myself. Mr Brilly has been gone from England a long time already; I have not seen the letter that you wrote, and I do not understand anything about that which you give me the honour of saying on the subject. I have not received any Paris Performance, except a short Novel entitled: *Disorders of the Civil War*, which the Author [Charles-Joseph Mayer] did me the honour of sending. I would become angry that the misfortunes which I myself had to endure should be the cause of so much trouble for you, if you were not recompensed by the approbation that your Work has received, which surely gives as much honour to your judgment as in the goodness of your heart. May you, Sir, always enjoy the honour and the esteem that you merit in all things. This is the sincere wish of she who has the honour to be with complete gratitude.

Sir,

Your very humble Servant, etc.

Theresa Asgill.

Le Barbier added: "(*) This letter never reached me. I have not had the honour of seeing the Gentleman to whom it was entrusted, and who surely has lost it during his travels."

He sent a copy of his play to Washington, who, in his response said: "for want of knowledge of the language, I can form no opinion of my own of the Dramatic performance Monsr Servitieur la Barbier."[447]

Other literary works appeared about Asgill during the course of his life, and after his death, most portraying him as a dashing young lover, falling for the most unexpected of females, such as Huddy's daughter, and in another, Betty, the daughter of a Pennsylvanian settler by the name of Penn! "... political doubts and divisions are eclipsed by the glorious spectacle of a young soldier, audacious in battle and noble in death, valiantly defending hearth and home from foreign attack. Similarly idealized portrayals of British military masculinity characterize the poetic tributes to John André, General Eliott and Charles Asgill that appeared during the 1780s. In the aftermath of a disastrous military campaign, consolatory depictions of the British officer as both martial hero and cultured man of feeling helped to obliterate the savagery of the conflict and to alleviate the bitterness of defeat."[448]

APPENDIX SEVEN

Asgill in Fiction

On 5 January 1813, the play, *Washington, Or The Reprisal* A Factual Drama, a play in three acts, in prose, was staged for the first time in Paris at the Théâtre de l'Impératrice. The play was written by Henri de Lacoste, Member of the Légion d'Honneur and l'Ordre impérial de la Réunion. In it we see Asgill fall in love with Betti Penn, the daughter of a Pennsylvanian Quaker, who supports him through his ordeal awaiting death. Henri de Lacoste's play does not appear to have had the same notoriety as the better known J. L. le Barbier-le-Jeune's *Asgill*.

American authors also thought his experiences were money-spinning opportunities. Their novels and plays took a lot of liberty with the truth; some having him marry American ladies, portraying him as a gallant Beau. One author had him marry a Loyalist schoolteacher named Madeline Burnham, in Trenton, New Jersey. They even depicted him as a very handsome fellow!

Asgill depicted in *Two Girls of Old New Jersey: A School-Girl Story of '76*, by Agnes Carr Sage, illustrated by Douglas John Connah. The image's caption reads "Clemency! For the British prisoner, your Excellency!"

Appendix Seven

In 1911 John Lawrence Lambe penned *Experiments in Play Writing, in Verse and Prose*, first published by Sir Isaac Pitman & Sons, London, Bath and New York, which is a collection of plays, one of which is *An English Gentleman, the story of The Asgill Affair retold*. In this very fanciful "re-telling," the "English Gentleman" is George Washington, and Asgill declares his love for Virginia Huddy (a name invented by the author), the daughter of the murdered Captain Joshua Huddy. The play ends with Washington's blessing on this union, when he says "Captain Asgill, it rejoices me that an unfortunate incident has terminated thus happily. (Taking his hand) May your union with this young lady symbolise the affection which I trust will ever unite the old country and the new. Sir, it has been your great happiness to win the best fortune of all, what is most adorable on earth – the love of a good and faithful woman."

What chance all the authors of books and plays thought Asgill might have had for romance, is an entirely other matter. It made for a good story though, and all of Europe had been rooting for Asgill to escape the gallows, so the plays and books became popular. In all, ten plays or books were written about Asgill, prior to Katherine Mayo's 1938 book, *General Washington's Dilemma*, and many more had substantial sections devoted to the events surrounding him in 1782. Seven of these were published during his lifetime. I have often wondered how he felt about this. While he was portrayed as a handsome and dashing young suitor, the so-called-facts were often a twisted version of the truth. It must have been difficult to live your life with these extraneous distractions, impacting your privacy and sensibilities. Did he ever watch any of the plays, I wonder?

Over the succeeding centuries, fiction became fact, very much to Asgill's detriment. The folklore became rumours attributed to Asgill. The fiction authors are in no small part responsible for the reputation suffered by their subject, through no fault of his own, for the past two centuries.

My cousin's wife, Kate Vaughan-Newton, kindly translated my copy of the Henri de Lacoste play. The original owner of this copy was an American, James Lorimer Graham (1835 – 1876). He was a French scholar, who immersed himself in the French way of life and French culture. He spent much of his life in France. It is clear from all the hand-written notations and newspaper clippings he inserted on the blank pages in the play, that the "Asgill Affair" fascinated him, perhaps with good reason since France was America's ally during the revolution. While the de Lacoste play was based on historical fact, great licence has been taken with the truth. I was, nevertheless, fascinated by this comment from J. L. Graham: "and every time, the mob, for some coins given to the soldiers, had the freedom of entry to him [Asgill], in order to insult him caged up, impatiently calling for the promised show, his execution." This is really interesting since this is exactly as Asgill himself described what happened to him during his incarceration at the tavern in Chatham. *Nowhere else have I ever seen this*, other than in Asgill's own letter which, in James Lorimer Graham's time, had never seen the light of day, publicly at least. How did he know this? Had he found the Letterbook before me?! Had he even been an owner, along the way to eventually languish on a Maggs bookshelf, awaiting a new owner?

What follows, below, is a transcription and translation of the extra notes and newspaper cuttings, written in French, inserted into the French play, *Washington, Or The Reprisal*. A friend of mine, Vicky Kennedy, at the time a student of French at the Open University, kindly translated the Graham additions, and this is what follows.

Page 71

Asgill (Charles, Sir), English General born towards the middle of the eighteenth century, died in 1823. He served at first in America, under the command of General Cornwallis: he had been taken prisoner in the siege of York-Town, in Virginia, and had been designated by fate to be executed in retaliation; later granted a pardon by the American Congress on the appeals of the French Government, following which he was sent back to England. He served in Flanders and in Ireland. This episode of Asgill's life has provided material for several plays in the theatre. [This has overlooked French author Charles-Joseph Mayer's 1784 novel, *Asgill, ou les désordres des guerres civiles,* which was the first piece to tell the story of 1782.]

Gentleman's Magazine, Vol XCIII, p. 274 – Gordon History of the Rise, ... and Establishment of the United States of America, Vol. IV p. 248

Page 73

The following article is a translation from the French newspaper, *Le Constitutionnel* of 22 October 1867, in which they quote from the *Mecure de France* of 3 November 1783:

Appendix Seven

The same day ... all Paris fought to get hold of the *French Mercury* in order to read the following: 'Captain Asgill arrived here with his mother and his two sisters. He had to go to Fontainebleu in order to thank The Count of Vergennes for his powerful intervention for which he owes his life. His sisters are very kind and the eldest especially is as beautiful as an angel. She is not yet fully recovered from the cruel attack of nerves she felt and which frightened her every day when she learned of the sad fate which threatened her brother.' Why so much emotion from us and what interest to Parisiennes is the comings and goings of an English family?

It is quite a long story and extremely strange which I will briefly summarise. It serves as a literary theme of this time, today entirely unknown, Monsieur de Mayer, in order to write in the Library of novels gathers periodically a headline which, it appears, had at that time a truly popular success *Asgill or the upheaval of Civil Wars*.

The author recalled the struggle of the American rebels against their British brothers and, in order to recount an episode of this war, a detail in short, and almost an anecdote, he started as follows:

I would be very tempted to go back to the first day of the world ... I would see the first human innocent and happy. And how was it for him? He held in his hand some flowers, some fruit and next to him an innocent and beautiful companion. Love which should only produce happiness had found a place in the cradle of the first man. This happiness was short lived. The first patriarchal dynasty etc.

Page 74

The gentleman finishes by passing through a great flood and, by a long detour, arrives at the year of grace 1782, where I hasten to give him the slip, having only quoted these ten lines in order to give you an idea of his work and his innocent manner to say the least.

The Charles Asgill Affair

Now, here are the facts:

Asgill, a young 20 year old Englishman, heir to a large fortune, left, in 1781, in order to fight on the other side of the Atlantic. At the first encounter, he was taken prisoner and taken to General Washington's camp. This *Fabius* received him with kindness and eased the harshness of his imprisonment. But, at this juncture, a dreadful attack which shamed humanity changed the situation and destroyed the hopes of the captive, who was going to be returned to London.

The English, after having seized a fort, in which all the garrison perished rather than surrender, had just hung, without any form of trial and on the nearest tree, Captain Huddy, of the Jersey Militia, one of the three brave survivors.

These examples of disloyalty and of barbarity were neither new nor unusual on their part. Washington resolved to retaliate. He started at once to write to Sir Henry Clinton, asking that Lippincott, the officer who had ordered the murder of Huddy, be handed over. The reply was late in coming; he called on fate to choose amongst the English prisoners a victim who would pay for the guilty, and fate marked out Asgill.

Page 75

As a consequence of this doleful lottery, drawn on towards the end of May, the hapless wretch spent almost six months between life and death, seeing without ceasing the gallows through the skylight of his dungeon, at a distance of 50 paces. Three times he had been taken, and every time, the mob, for some coins given to the soldiers, had the freedom of entry to him, in order to insult him caged up, impatiently calling for the promised show, his execution.

But his family and his friends, and, on both sides of the Atlantic, foreigners, were upset for him, and worked to save him. His mother, with his sisters, went to throw themselves at the feet of the King and Queen of

Appendix Seven

England; she obtained a positive order to deliver up Lippincott. But will the ship arrive in time? If it was captured en route! Fear nothing, said George III, the captain will be told the object of his mission, and the fair-minded Louis, who allowed Cook to navigate in safety, will still respect my flag.

In London, it was the sole conversation and, according to custom, the betting opened in this connection on the harbour, in the markets and the taverns; the guineas rolled, one fought each other, one beat each other, one aimed a punch, one became drunk on beer while waiting for the news.

The Dutch offered their services; they sent a letter to Washington and on the 1st June, the innocent person had already stepped onto the fatal ladder, when their message granted a stay of execution.

It was postponed a second time on the authority of the Marquis de Rochambeau; without being informed of the intentions of his sovereign, he personally took the initiative of this kind intervention.

Page 76

Finally, for the third time, as the unhappy young man, brought back in front of the gallows, had the rope around his neck, a horseman arrived, flat out and screaming; Mercy! Mercy!

He brought an order from Congress, an order to free the condemned man, a special letter from Washington, and the copy of a dispatch signed Vergennes, dated Versailles, 29 July. (It was then in November).

Lady Asgill had finally appealed and immediately obtained the intervention which had been decisive, to the King of France. The Count of Vergennes had immediately dispatched to General Washington a very good letter, but too long for it to be reproduced here.

The Charles Asgill Affair

The source from where I extracted all these details, the order in full, with the warm thanks that Lady Asgill addressed at once to our Ministry of Exterior Relations.

According to her son, on returning to Europe, the following year, he became, in Paris and Versailles, the lion of the autumn season. He rushed to Fontainbleau to show to M. de Vergennes his gratitude, and the Court like the town celebrated his deliverance; What he says! Writing to each other on this subject the good M. de Mayer, who he says welcomes him like the Queen, the King and his dignified Minister have done! He will never dare to arm himself against a nation, to whom the life of a man is so precious.

On this particular question, neither the English officer nor his leader did not even try to attempt a response for the moment; perhaps they were highly embarrassed; the events took care of themselves.

Between Great Britain, the United States and France, hostilities were suspended from the month of January, and peace was signed in Paris on 30th November.

The above article is cut from the *Constitutional* newspaper of 22 Oct 1867. For other paragraphs on the Asgill Affair see the *Mercury* for 1782, (July, p. 173. Aug pp 10, 61, 142, 159, 205; September pp 15, 47, 116, 196 and for 1783 January pp 17, 143)

Page 83

In the 'Extract from the Life of M. de Vergennes' which accompanies by way of a preface to the *'Historical and Political Report on the reign of Louis'* one reads the following:

Sparing with compliments, he was lavish with kindness; misfortune was the first title of his charity: his behaviour towards the mother of Asgill will arouse in all times the gratitude and the sensitivity of every mother. In the war between America and Great Britain, Lippincott, an English Officer, massacred in cold blood.

Appendix Seven

Page 84

Huddy, an American officer, whom he had taken prisoner. Lippincott himself escaped. It is in vain that M. Washington asks that one delivers to him the guilty person.

Congress orders that all the English Officers held as prisoners must draw lots, and that, for atonement, one of them will lose his life. The young Asgill is the one whom fate marks out; he is condemned, but he has not yet suffered torture; Washington delays his execution.

The mother of this unfortunate asks again to everyone in power; she writes a letter

Page 85

to M. de Vergennes, a letter of heartbreak; the leading minister conveys this to the Queen; she was a mother; tears flow from her eyes and those of the minister; they strike to the heart of Louis XVI: the Monarch writes to Washington. Asgill is saved. Soon M. de Vergennes has the warm satisfaction of receiving the thanks and the blessings from a mother and a son who have both been given a second life.

APPENDIX EIGHT

Sophia Asgill's letter to Thomas Graham

The only letter written by Sophia Asgill to Thomas Graham to survive is the one she wrote on 17 May 1799.[449] It is transcribed here in full, courtesy of the National Library of Scotland which holds the original. Graham had been at Messina in Italy just two months when it was written. In their secret pact they undertook not to top or tail their letters.

After Sophia's death, doubtless to Lynedoch's astonishment, all those letters of his, which she had kept with so much care, were found amongst her effects, and were returned to Lynedoch. They formed a treasure trove of information about the Peninsular War hero; historians should be glad that Sophia did not keep her part of the bargain.

Berkeley Square, May 17th, 1799[450]

Colonel Hope sends me word that if I will instantly write to you he can get my letter conveyed safely. I seize a pen, therefore, without a moment's delay, and am happy in the opportunity of giving you some account of myself and friends – It is about a month I think since I wrote last when Cox and Greenwood took charge of my letter, and not Sir Gordon as I told you – I am now very impatient to hear from you. Your letter of March 1st has been acknowledged, and we have not heard of you since, – but remember we don't forget you, and I, for one, never shall. This I promise you.[451] – Bab[452] arrived in Town on Tuesday, looking remarkably well, and feeling rather rejoiced at coming amongst us again, for now Bratty[453] is at Brighton, the charms of Farnbro' are greatly diminished. We have been incessantly together since her arrival – yesterday she dined tête à tête with me, and in the evening Mr Streat'd[454] joined our party and we were a most comfortable trio – he remains in Town a few days longer – his spirits are as usual, good – and in all aspects I think he remains exactly in the same state as when you saw him – I have never known as many Balls, Dinners, and Assemblies given as this year. I am quite sick of them, but am not tired of living in Berkeley Square and seeing several persons I love, almost every day. These,

Appendix Eight

however, are of my own sex, therefore I hope I may be indulged with their society.

Mrs P and V[455] are well, the former I fear is destined this Summer for Ireland, the latter as I believe I mentioned before, for Scotland – very lately I have heard from Ireland, and as I am desired to send over a cargo of wearing apparel I conclude we have little chance of their return to this Country. She says they are all expecting the French fleet and are in readiness for them, it is now however generally believed here, that they are gone to join the Spanish fleet – She does not write in good spirits yet she makes very few complaints – it is evident St. C:[456] does not encourage any wish she may entertain of a visit to her friends, nor do I think he would apprehend danger for her even tho' there should be cause for it. I feel sad and sorry but can do nothing. On Tuesday the Poles, Villiers's, G. Seymours and Ourselves dined with Colonel Hope; a Bachelor, and Bachelor's House, are most charming things, not to mention a Bachelor's Dinner – He has furnished his House in the best taste and in the most comfortable manner you can imagine, and to be short with you – the four Ladies in question were quite charmed with their Landlord. What a happy mind that man has! He is always gay and unrepining. – I often wonder to see him so cheerful while he is constantly feeling so much inconvenience from his lameness.

The Sheridans are in town. Next week the long-expected new play called Pizarro, (translated from the German of Kotzebue and the story from Marmoutel's Incas) is to come out.[457] It was originally illtranslated by some one who sent it to Sheridan, and he has completely written it over again, and has altered it and added to it according to his fancy. They say it will be by much the best thing that has appeared for years, and the world are in high expectation. But like the day it comes out at the end of the Season, owing to his idleness. Poor Hester [her cousin, Hester Ogle, and Sheridan's second wife] has suffered agonies of mind about this Play, and continues as violently attached to him as ever. She never goes out, but waits and watches at home with her accustomed vigilance –

Henry[458] is a Student in the Temple except when he relaxes from business in Hertford St. We have not heard from my eldest Brother lately [this is Charles now, since George and

Edward had died at sea prior to this letter being written]. My Father wishes him to come home and he is to do as soon as he possibly can. Mrs [Barbarina] Wilmot does not give a good account of my Father's state of health, therefore thinks it encumbent [sic] on him to return on this account. I know very well you will not fail to let us hear from you whenever you can, and therefore it is unnecessary to remind you of it – no one understands the comfort of one line only from an absent friend so well as you. I am sure of this from your actions.

APPENDIX NINE

"The Swindler Asgill" Press Reports

From the following excerpts, from half a dozen newspaper reports on "The Swindler Asgill," what little is known emerges. Focus will be placed on the similarities which exist between the back-story of William Charles Asgill, and what was declared about the Swindler. The similarities form a clear link. The italicized text is directly quoted, and the plain text gives an assessment of those quotations.

Devizes and Wiltshire Gazette, 4 September 1823: *Swindler, a Gentleman has for Some Months.* This newspaper is the first known account – and this may be why it is the only one that does not mention the name "Asgill" in connection with "The Swindler." Nevertheless, the story is exactly the same as all the others printed during September and October 1823. This newspaper followed up with a fuller account, mentioning the name "Asgill," on 18 September 1823 (see below).

1) *The title of Baronet or not.* This suggests to readers that he *might* become a Baronet, reflecting William Charles Asgill's "concocted story" of one day being the 3rd Baronet Asgill.

2) *He was represented to be the nephew of a baronet, and heir to the title and estates.* The William Charles Asgill family-story could so easily have changed over the decades to fit with William's obituary which stated that he was a son rather than a nephew of the General.

3) *this presumptive Baronet.* He was just that – attempting to con people into believing he was something he was not.

4) *and the titled youth turns out to be a noted Swindler.* The simple fact – unpleasant though it is.

5) *and his Chere Amie a confederate through whom he transmitted his plunder to the gang in London.* This might refer to Mary Luetchford, since he married her the following month, on 9 October 1823.

6) *this same gentleman has since visited Bath, where he has proved equally liberal in patronizing the tradesmen. Several instances have come to our knowledge of his having contracted debts to large amounts, all of which, as in Dorsetshire, he has left unpaid.* Clearly William Charles Asgill moved about in order to remove himself from

The Charles Asgill Affair

locations where people were already suspicious, thus giving him a wider field within which to operate.

Southern Reporter and Cork Commercial Courier, 13 September 1823 (*Extraordinary* Adventurer)

1) *The answer was decisive – Sir Charles had no nephew.* So said General Sir Charles Asgill's Executors. The Swindler had been found out in his lie about being related to the General. Perhaps that is why he decamped to Bath, or Salisbury, or Winchester, or Durham, or Devizes, or even Cork, the locations where articles appeared in the press.

2) *let fall some hints about a certain valuable living which would be in his gift and expressed a doubt whether he would take the title of Baronet or not.* Hence the story William Charles Asgill told about being the legitimate son of General Sir Charles Asgill who was so cruelly disinherited. In reality, had William, or any other legitimate son of the General actually been disinherited, only an Act of Parliament could have denied him the Baronetcy. Had that happened then this would have been revealed in the course of all the research done by so many people over the past 100 to 150 years. Not to mention the College of Arms stating categorically that the General died without issue.

Devizes and Wiltshire Gazette, 18 September 1823 (The Shaftsbury Adventurer)

1) *talking familiarly of Eton and Oxford.* Perhaps, whoever William Charles Asgill really was, he *had* actually been educated at Eton and Oxford? In any event, one of the newspapers mentioned that he had been Head Boy at a public school. If one established who were the Head Boys at Eton at the time William was about 16 (circa 1814-1816), then maybe one of those surnames might be found on Robert Asgill's DNA results?[459] Another line of enquiry might be possible if records of the "Society for the protection of trade against swindlers and sharpers – established March 25th 1776" could be found.

2) *well educated, and smart and agreeable in conversation.* These are all attributes to become a school teacher, the profession chosen by William Charles Asgill, who had his own school in Beddington, Surrey, after the events of 1823. The same attributes were also useful to a con-artist.

Durham County Advertiser, 20 September 1823 (*Extraordinary Swindler*)

Appendix Nine

1) *aged about five or six and twenty.* So the Swindler would have been born in approximately 1798 (according to guesswork in this article) which ties in perfectly with William's age – he is believed to have been born circa 1800, according to his death certificate. No baptism record has ever been found, only census returns for later in his life. For a birth certificate to be found we would need to know who he really was at birth, and it certainly wasn't *"William Charles Asgill"* who was supposedly born in Ewell, Surrey, circa 1800. If this man was born in Ewell, by what name was his baptism registered one wonders?

2) *exhibited himself to the townspeople in the full regimentals of the Guards.* As part of his disguise as a nephew of the General, the Swindler portrayed himself as being an army officer – just like his "uncle" General Sir Charles Asgill – even choosing the regimental uniform of the 1st Regiment of Foot Guards. Also found was a newspaper article referring to Major and Mrs Asgill (*Morning Post* (London), 11 December 1823) – William had married Mary Leutchford the previous October, soon after the General had died, so this could well have been William and Mary Asgill (a.k.a. The Swindler Asgill and his *"cher ami"*).

3) *whether he would take the title of Baronet or not.* There is no choice about such things – the baronetcy would have continued willy-nilly but could only have been conferred on a legitimate son of General Sir Charles Asgill. The General had no legitimate heirs.

4) *There is good reason to believe that the real name of "Mr Asgill" has been discovered and that it is not altogether unknown in the annals of the police, but for obvious reasons we omit it for the present.* One newspaper mentioned the name "Armitage," but proof would be needed to establish that it was him.

5) *in person about the middle size, high shouldered, rather well looking, with fine dark eyes.* He seems to have been good looking, as well as personable and well educated.

Salisbury and Winchester Journal, 22 September 1823 (*The Swindler Asgill*)

1) *The farce was kept up by a list of books being ordered, and the embryo of a Divine duly attending in the study every morning.* William Charles Asgill was training to be a clergyman. The story that the General wanted to disinherit William for *not* wanting to be a clergyman is rather ridiculous in itself. The priesthood was a very unlikely path for the first

The Charles Asgill Affair

born son of a general and of a baronet; perhaps this is why, by the time William Charles Asgill died, his obituary declared that he was "the second son of Sir Charles Asgill." A second-born son would have fewer expectations on him. The natural course for a first-born would have been to join the army. And to disinherit your son because he didn't want to be a clergyman simply doesn't ring true.

2) *and wishing to procure his ordination without the formality of a University education.* Here is the origin of the story of William Charles Asgill training to become a clergyman – one of the main similarities between him and The Swindler Asgill. "Without the formality of a University education" also doesn't fit with the claim that he was a graduate.

3) *claimed for his cousin a Baronet, whose name he has lately honored by assuming.* He took the Asgill name without any formal or biological connection to the family. This could hardly be described as an "honour" from General Sir Charles Asgill's lens.

4) *a nephew, ... a young man of superior attainments, gentlemanly manners, and amiable disposition, ... who had received a highly finished education.* Thus giving him every opportunity and qualification to become a schoolteacher – or a con-man.

5) *At an early age he distinguished himself as the head boy of a public school, from which he was removed at the age of sixteen to one of our Universities.* Once again, stressing well educated. This education stood him in good stead to mastermind complex scams, and also enabled him to become a schoolmaster.

6) *ill conduct have induced his family for some time to disown him.* This is akin to the disinheritance story. The subtle difference between "disown" and "disinherit" can become lost in the telling over the decades; "his family," whoever they may be, are the ancestors of the present day descendants of William Charles Asgill. But who was he, really? What was his real name?

Morning Post (London) 11 December 1823

Fashionables In Town – Major & Mrs Asgill. This looks suspiciously like William and his new wife, Mary (his *cher ami*), two months after they married. William Charles Asgill had an army officer's uniform made for him (without paying for it) and was letting it be known he had been in the Guards. Being amongst distinguished company was

Appendix Nine

clearly what The Swindler enjoyed, and afforded him credence. It looks very much like "hiding in clear sight," since there is no other contender, to my knowledge, who had gone by the title "Major Asgill" in 1823.

Morning Chronicle (London), 14 June 1834

Fashionable Swindlers – More than 300 men were known swindlers in London by 1834. So swindling was very commonplace. This may well be after William's time as a swindler, but even so there are a number of articles relating to other people doing the same thing in 1823. On the other hand, if William Charles Asgill was finding that his school was not as profitable as he needed it to be, he could have turned to swindling and continued to get away with it. This might be why he found himself in the Debtors' Jail by 1832. If the charge against him then could be established, more would very likely follow. Finding out who William Charles Asgill really was is the only way to further progress the ancestry of the present-day Asgills. It really looks as though William *did* get away with it, unpunished. The only possibility that he didn't is his prison term in the Debtors' Jail – if the sentence really did cover his activities of 1823. If we knew William's birth surname a very exciting family history could well fall into place. But the conclusion is that the only chance of this happening now is to find a DNA specialist; one able to find a man with no name and no date of birth, and only the belief that "he was born in Ewell circa 1800."

A very tall order, to put it mildly.

The Charles Asgill Affair

BIBLIOGRAPHY

1851 census, HO 107-1616. The National Archives.

A Family Chronicle derived from Notes and Letters selected by Barbarina, the Hon. Lady Grey. Edited by Gertrude Lyster. London: John Murray, 2020.

A list of the governors of St. Thomas's-Hospital, in Southwark, 1768. London: St. Thomas's Hospital, 1768.

A List of the Officers of the Army, 1783. WO 65/33. The National Archives.

Abel, Martha. "'Unfortunate': Lancaster, Pennsylvania, May 26–28, 1782." *The Journal of Lancaster County's Historical Society* Vol. 120 No. 3 (2019): 97–105.

Acronym Definition. https://acronyms.thefreedictionary.com.

Ammundsen, Anne. "Miss Asgill's Minuet." *Metropolitan: The Journal of the London Westminster & Middlesex Family History Society* Vol. 7 No. 2 (2021): 88–89.

Ammundsen, Anne. "Saving Captain Asgill." *History Today*, Vol. 61 No. 12 (December 2011): 38-43.

Aris's Birmingham Gazette, 1782.

Asgill, Charles. Letter to the Editor of the *New Haven Gazette*, 20 December 1786. Private collection.

Asgill, Charles. Letter to an unidentified recipient, 9 February 1821. Morgan Library, New York.

Asgill Pedigree, 1830. Derbyshire Record Office.

Aspinall-Oglander, Cecil Faber. *Freshly remembered: the story of Thomas Graham, Lord Lynedoch*. London: Hogarth Press, 1956.

Bath Chronicle, 1782, 1788, 1823.

Blackburn Standard, 1854.

Brett-James, Antony. *General Graham*. London: Macmillan, 1959.

British Headquarters Papers. PRO 30. The National Archives.

Brown, Steve. British Regiments and the Men Who Led Them 1793-1815: 1st Regiment of Foot Guards. https://www.napoleon-series.org/military-info/organization/Britain/Infantry/Regiments/c_1stFootGuards.html.

Bryant, Arthur. *Years of Victory 1802-1812*. London: Collins, 1944.

Burton, William E. *Cyclopedia of Wit and Humor, of America, Ireland, Scotland, and England.* New York: D. Appleton & Co., 1858.

Butler, Marilyn. *The Works of Maria Edgeworth.* Part I. Routledge, 2019.

Byron's "Corbeau Blanc" The Life and Letters of Lady Melbourne. Edited by Jonathan David Gross. Houston, TX: Rice University Press, 1997.

Calcutta Gazette, 1784.

Cannon, Richard. *Historical Record of the 11th Foot or North Devon Regiment of Foot.* London: Parker, Furnivall and Parker, 1845.

Chambers, Liam. *The 1798 rebellion in north Leinster.* Dublin: Four Courts Press, 2003.

Colonial and State Records of North Carolina Vol. 17. Raleigh, NC: Josephus Daniels, 1899.

Colonial Office Papers, The National Archives.

Colvile, Lady, and Zelie Isabelle Richaud de Preville Colvile. *History of the Colvile Family.* Edinburg: Morison and Gibb, 1896.

Correspondence of Sir Thomas Graham. National Library of Scotland.

Counihan, Alan. "A Cautionary Tail." *Local History Review* Vol. 18 (2013), pp. 87–90.

Curtis, Edward R. *The Organization of the British Army in the American Revolution.* Gansvoort, NY: Corner House Historical Publications, 1998.

Curzon, Catherine. *The Scandal of George III's Court.* Barnsley, UK: Pen & Sword Military, 2018.

Debrett's Baronetage of England. London: William Pickering, 1840.

Derby Mercury, 1783.

England and Wales, Civil Registration Index: 1837-1983 Record, Maidstone, Kent.

Ewald, Johann. *Diary of the American War: A Hessian Journal.* Edited by Joseph P. Tustin. New Haven, CT: Yale University Press, 1979.

Founders Online, National Archives, https://founders.archives.gov.

Freemans Journal (Dublin), 1807, 1812.

Garden, Alexander. *Anecdotes of the American Revolution,* second series. Charleston, SC: A. E. Miller, 1828.

Geoghegan, Patrick. *Robert Emmet: A Life*. Montreal: McGill-Queen's University Press, 2002.

Geograph.org.uk.

"George Montagu, 4th Duke of Manchester papers 1779-1788," William L. Clements Library.

Georgian Papers Online, Royal Collection Trust.

Gloucester Journal, 1840.

Gower, Granville Leveson. *Lord Granville Leveson Gower (first earl Granville): private correspondence, 1781 to 1821*. Edited by Castalia, countess Granville. Volume 1. New York: E. P. Dutton, 1916.

Graham, Samuel. *Memoir of General Graham*. Edited by Colonel James J. Graham. Edinburgh: R. & R. Clark, 1862.

Grant, Philip R. *A Peer Among Princes*. Barnsley, UK: Pen & Sword Military, 2019.

Hagist, Don N. "Unpublished writings of Roger Lamb: Soldier in the American War of Independence." *Journal of the Society for Army Historical Research* 90, no. 362 (Summer 2012): pp. 77-90.

"Hale, Edward Everett (1822-1909)." Harvard Square Library.

Hale, Edward Everett. *The life of George Washington, studied anew*. New York and London: G.P. Putnam's Sons, 1888.

Hamilton, Sir Frederick William. *The Origin and History of the First or Grenadier Guards*, 3 Vols. London: J. Murray, 1874.

Hamilton, Henry. "The Failure of the Ayr Bank, 1772." *The Economic History Review*, Vol. 8, No. 3 (1956), pp. 405–17.

Hampshire Local Studies Collection, Hampshire Record Office.

Handbook of the Antiquarian Booksellers Association 2008-9. Edited by Michael Silverman.

Harris, Cicero W. "Early Times in Granville County." Transcribed by Tina Tarlton Smith. *The Torch Light*, 19 March 1878.

Henriques, Peter R. "'Unfortunate': The Asgill Affair and George Washington's Self-Created Dilemma." First and Always: A New Portrait of George Washington. Charlottesville, VA: University of Virginia Press, 2020.

Hereford Journal, 1783.

Hibernian Journal, 1782.

Hilton Price, F. G. *A Handbook of London Bankers, with some account of their predecessors the Early Goldsmiths.* London: The Leadenhall Press, 1891.

"History," All Saint's Church, Loose, UK.

Hoare, Stephen. *Palaces of Power: The Birth and Evolution of London's Clubland.* Cheltenham, UK: History Press, 2019.

Insured: Mary Ann Mansell, 15 Park Place South near The Man in the Moon Chelsea, MS 11936/488/978737, London Metropolitan City Archives.

"Ira Lipman, Security Man Who Spoke Out for Air Safety, Dies at 78." *The New York Times*, 29 September 2019.

Ipswich Journal, 1785.

James, Charles. *A collection of the charges, opinions and sentences of general courts martial.* London: T. Egerton, 1820.

Jones, T. Cole. *Prisoners of War and the Politics of Vengeance in the American Revolution.* University of Pennsylvania Press, 2019.

Jones, Thomas. *History of New York During the Revolutionary War: And of the Leading Events in the Other Colonies at that Period.* Volume. 2. New York: The New York Historical Society, 1879.

Kelly's Court Directory, 1866-1867.

Kentish Gazette, 1790, 1870.

Kidd's Own Journal (London), 1852.

Kosmetatos, Paul. *The British Credit Crisis of 1772-3.* Edinburgh: University of Edinburgh, 2018.

Lamb, Roger. *An original and authentic journal of occurences during the late American war, from its commencement to the year 1783.* Dublin: Wilkinson & Courtney, 1809.

Ledger of Accounts of the 1st Company, 1st Battalion, Brigade of Guards, November 1777 - March 1779, National Archives and Records Administration, Washington, DC.

Leeds Intelligencer, 1785.

"Letters of Charles O'Hara to the Duke of Grafton." Edited by George C. Rogers, Jr. *South Carolina Historical Magazine*, v. 65 (1965): 158-80.

Lever, Charles. *Jack Hinton, The Guardsman.* London, George Routledge & Sons Ltd., 1843.

London Daily News, 1903.

Masters, Brian. *Georgiana Duchess of Devonshire*. London: Hamish Hamilton, 1981.

Mayo, Katherine. *General Washington's Dilemma*. London: Jonathan Cape, 1938.

McKay, William and W. Roberts, *John Hoppner RA*. London: P. & D. Colnaghi & Co and George Bell and Sons, 1909.

Montgomery, Robert. *The Hawks of Hawk-Hollow. A Tradition of Pennsylvania*. Philadelphia: Carey, Lee & Blanchard, 1835.

Moore, Pam. "Sir Charles Ogle: A Worthy Admiral." *The Portsmouth Papers*, 53 (1988).

Morgan, Lady (Sydney) nee Owenson. *The Book Of The Boudoir*. Volume 1. New York: Harper, 1829.

Morning Post (London), 1823.

Musgrave, Richard. *Memoirs of the different rebellions in Ireland*. Dublin: Robert Parchbank, 1801.

Muster book of HMS *Warwick*, ADM 36/8657, The National Archives.

New Times (London), 1823.

New York Gazette, 1781.

Northampton Mercury, 1782.

Norwich Packet, 1798.

Ogle, Charles. Letter to Thomas Phillips, 23 October 1823. Morgan Library, New York.

Old Bailey Proceedings Online. www.oldbaileyonline.org.

Office-Holders in Modern Britain: Volume 11 (Revised), Court Officers, 1660-1837. Edited by R.O. Bucholz. London: University of London, 2006.

Orders and Instructions for His Majesty's 33d Regiment of Foot on their embarking for North America, 29 November 1775. WO 26/29 pp. 156-157. The National Archives.

Orders for the 2d Battn of the 71st Regt on its embarkation for North America, 25 April 1776. Concord Free Public Library, Concord, MA.

Oxford Dictionary of National Biography. Oxford: Oxford University Press, 2004.

Oxford Journal, 28 November 1812.

Pakenham, Thomas. *The Year of Liberty*. London: Hodder and Stoughton, 1969.

Parliamentary Papers. Volume 36. London: HM Stationers, 1854.

Pearse, Hugh Wodehouse. *Memoir of the life and military services of Viscount Lake: Baron Lake of Delhi and Laswaree, 1744-1808*. Edinburg: W. Blackwood and Sons, 1908.

Philhower, Charles A. *Brief history of Chatham Morris County, New Jersey*. New York: Lewis Historical Publishing Co, 1914.

Philippart, John. *The Royal Military Calendar*. Volume I. London: J. Valpy, 1815.

Philippart, John. *The Royal Military Calendar*. Volume II. London: J. Valpy, 1820.

Phillimore, Robert. *Commentaries upon International Law*, Vol. 3. T & J.W. Johnson & Co., 1857.

Pierce, Arthur D. *Cradles of Revolt: The Taverns*. New Brunswick, NJ: Rutgers University Press, 1988.

"Poor Mary Maher." *The Celt* (Dublin), 28 November 1857.

Price, F. G. Hilton. *A Handbook of London Bankers, with some account of their predecessors the Early Goldsmiths*. London: The Leadenhall Press, 1891.

Proceedings of the New Jersey Historical Society. Volume 8, 1856-1859. Newark, NJ: printed at the Daily Advertiser Office, 1859.

Raftis, Edmund B. *Summit, New Jersey: From Poverty Hill to the Hill City*. Carolina, RI: Great Swamp Press, 1996.

Reading Mercury, 1782.

Records created, acquired, and inherited by Chancery, and also of the Wardrobe, Royal Household, Exchequer and various commissions. The National Archives.

Records of the Prerogative Court of Canterbury. The National Archives.

"Remarkable Domestic Events, August 1791." *The Historical Magazine,* Vol. 3 No. 34 (August 1791).

Robertson of Kindeace papers, National Archives of Scotland.

Robinson, R.E.R. *The Bloody Eleventh: History of the Devonshire Regiment*. Vol. 1, 1685-1815. Exeter: Devonshire and Dorset Regiment, 1988.

Saunders's News-letter (London), 1782.

Shaw, William A. *The Knights of England*. Volume 1. London: Sherratt and Hughes, 1906.

Shelley, Lady Frances (Winckley). *The Diary of Frances Lady Shelley 1787–1817.* Volume 1. Edited by Richard Edgcumbe. New York: C. Scribner's, 1912.

Sichel, Walter. *Sheridan: from New and Original Material; Including a Manuscript Diary by Georgiana Duchess of Devonshire.* London: Constable & company ltd., 1909.

Simes, Thomas. *A Military Guide for Young Officers.* London, 1776.

Simes, Thomas. *The Regulator.* London, 1780.

Sir Henry Clinton Papers. William L. Clements library, Ann Arbor, Michigan.

"Sliabh na mban – Slievenamon." https://irishpage.com/songs/slevmoan.html.

Smith, Digby. *The Napoleonic Wars Data Book.* Barnsley, UK: Greenhill, 1998.

Statement of the Service of Lieutenant General Sir Charles Asgill Bart. Colonel of the 11th Regiment of Foot. WO 25/744 A, pp. 8-12. The National Archives, Kew, UK.

Supplementary Despatches, Correspondence, and Memoranda of Field Marshal Arthur Duke of Wellington, K.G. Volume 6. London: John Murray, 1860.

Survey of London: Volumes 29 and 30, St James Westminster, Part 1. Edited by F H W Sheppard. London, 1960.

Survey of London: Volumes 31 and 32, St James Westminster, Part 2. Edited by F H W Sheppard. London, 1963.

Tarleton, Banastre. *A History of the Campaigns of 1780 and 1781, in the Southern Provinces of North America.* Dublin: Colles, Exshaw, White, H. Whitestone, Burton, Byrne, Moore, Jones and Dornin, 1787.

Tarling, Barbara Frances. *Representations of the American War of Independence in the late eighteenth-century English novel.* PhD thesis, 2010, The Open University.

Taylor, James. *The Great Historic Families of Scotland.* London: J. S. Virtue, 1887.

The American Rebellion: Sir Henry Clinton's Narrative of His Campaigns, 1775-1782. Edited by William B. Willcox. New Haven, CT: Yale University Press, 1954.

The American Revolution in North Carolina, https://www.carolana.com/NC/Revolution/.

The Daily Commonwealth (Topeka, Kansas), 1887.

The Charles Asgill Affair

The Dispatches of Field Marshal the Duke of Wellington. Edited by John Gurwood. Volume 7. London: John Murray, 1852.

The Examiner (London), 1813.

The Edgeworth Papers, National Library of Ireland.

The Gentleman's Magazine, 1772, 1788, 1823.

The History of Parliament: the House of Commons 1754-1790. Edited by L. Namier. Martlesham, Suffolk: Boydell & Brewer, 1964.

The Life, Public Services, Addresses and Letters of Elias Boudinot, LL.D, President of the Continental Congress. Edited by J.J. Boudinot. Volume I. Boston and New York: Houghton, Mifflin and Company, 1896.

The London Chronicle, 1782.

The London Gazette, 1783, 1798, 1800, 1805, 1806, 1807, 1814, *1823.*

The New Annual Register for the Year 1790.

The New Haven Gazette and Connecticut Magazine, 1786.

The New York Gazette and Weekly Mercury, 1783

The New York Packet, 1782.

The Observer, 1938.

The Scots Magazine, 1788.

The Times (London), 1785, 1788, 1791, 1794, 1795, 1802, 1815, 1839.

The Writings of George Washington. Edited by W. C. Ford. Volume X, (1782-1785). New York: G. P. Putnam's Sons, 1891.

The Writings of George Washington from the Original Manuscript Sources 1745-1799, Vol. 24-25. Edited by John C. Fitzpatrick. Washington, DC: Government Printing Office, 1938.

Thompson's Compleat Collection of 200 Favourite Country Dances, Vol. 3. London: Saml. Ann & Peter Thompson, 1773.

Tighe, Mary. "The Shawl's Petition, to Lady Asgill. 1772-1810." *Psyche, With Other Poems.* London: Longman, Hurst, Rees, Orme, and Brown, Paternoster Row, 1811.

"Traditions of the American War of Independence." *United Service Magazine and Naval Military Journal* Part 3 (November 1834), 309-23.

Trial of Richard Lippencott. WO 71/95 p. 320-408. The National Archives.

Turner, George and James Russell. *Reports of Cases Argued and Determined in the High Court of Chancery*. London, 1823.

Turtle Bunbury Irish Histories. The Magistrate: Benjamin Bunbury (1751-1823) of Moyle & Killerig. https://turtlebunbury.com/document/benjamin-bunbury-magistrate/.

Vanderpoel, Ambrose. *History of Chatham New Jersey*. New York: Charles Francis, 1921.

Voice of Rebellion - Carlow in 1798: The Autobiography of William Farrell. Edited by Roger Joseph McHugh. Dublin: Browne and Nolan, 1949.

Wallace, Robert Hugh. "Royal Downshire Militia. Extracts from Order Books, &c. (Continued)." *Ulster Journal of Archaeology*, Vol. 13, No. 2 (1907); pp. 59–69.

Ward, Matthew H. *Joshua Huddy: The Scourge of New Jersey Loyalists*. The Journal of the American Revolution, October 8, 2018, https://allthingsliberty.com/2018/10/joshua-huddy-the-scourge-of-new-jersey-loyalists/.

Westminster School archives.

Wheeler, John H. *Reminiscences and memoirs of North Carolina and eminent North Carolinians*. Columbus, OH: Columbus print works, 1884.

Wilts & Gloucestershire Standard, 1845.

Women's Literary Networks and Romanticism A Tribe of Authoresses. Edited by Andrew O. Winckles and Angela Rehbein. Liverpool: Liverpool University Press, 2017.

The Charles Asgill Affair

NOTES

1 "The clearest evidence of the absolute necessity of self-preservation is required to palliate any infraction of the rights of a prisoner of war. The prisoner who has yielded under conditions, cannot be injured so long as he fulfils his part of the condition." Robert Phillimore, Commentaries upon International Law, Vol. 3 (T & J.W. Johnson & Co., 1857), p. 145.

2 Articles of Capitulation between Washington and Cornwallis, 19 October 1781, *Founders Online*, National Archives, https://founders.archives.gov/documents/Washington/99-01-02-07199.

3 One of the British officers present later identified the commissary of prisoners as "Mr. Witz"; the man's identity has not been confirmed. Samuel Graham, *Memoir of General Graham*, ed. Colonel James J. Graham (Edinburgh: R. & R. Clark, 1862), p. 32. And in a footnote on page 477 of *Biographical History of Lancaster County Pennsylvania* by Alexander Harris (Genealogical Publishing Co. Inc. Baltimore, 1977) the Dragoon Officer is named as "Captain Stake" and the commissary of prisoners as "Colonel George Gibson". These anomalies cannot be confirmed.

4 Katherine Mayo, *General Washington's Dilemma* (London: Jonathan Cape, 1938), pp. 263-68. This letter does not appear in the New York: Harcourt, Brace edition.

5 William E. Burton, *Cyclopedia of Wit and Humor, of America, Ireland, Scotland, and England* (New York: D. Appleton & Co., 1858), 1067.

6 Mayo, *General Washington's Dilemma*, p. 131.

7 Ibid., pp. 263-68.

8 Asgill Pedigree, 1830, Derbyshire Record Office.

9 *The Scots Magazine*, 1 September 1788.

10 F. G. Hilton Price, *A Handbook of London Bankers, with some account of their predecessors the Early Goldsmiths* (London: The Leadenhall Press, 1891), p. 123.

11 *A list of the governors of St. Thomas's-Hospital, in Southwark, 1768* (London: St. Thomas's Hospital, 1768).

12 *The Gentleman's Magazine and Historical Chronicle*, September 1788, p. 841.

13 *Oxford Journal*, 28 November 1812.

14 Paul Kosmetatos, *The British Credit Crisis of 1772-3* (Edinburgh: University of Edinburgh, 2018).

15 *The Gentleman's Magazine*, June 1772, p. 293.

16 Henry Hamilton, "The Failure of the Ayr Bank, 1772," *The Economic History Review*, Vol. 8, No. 3 (1956), pp. 405–17.

17 "Asgill's Rant," *Thompson's Compleat Collection of 200 Favourite Country Dances*, Vol. 3 (Saml. Ann & Peter Thompson, London, 1773). And on "Charles Asgill's" YouTube.

18 *The Gentleman's Magazine*, September 1788, p. 841.

19 Mayo, *General Washington's Dilemma*, p. 162.

20 Ibid., p. 161.

21 Alexander Garden, *Anecdotes of the American Revolution*, second series (Charleston, SC: A. E. Miller, 1828), pp. 33.

22 Mayo, *General Washington's Dilemma*, p. 163.

23 Autograph of Charles Asgill, Autograph Book, University of Göttingen Archives, https://commons.wikimedia.org/wiki/File:Charles_Asgill_autograph_book_1778.jpg.

24 Statement of the Service of Lieutenant General Sir Charles Asgill Bart. Colonel of the 11th Regiment of Foot, WO 25/744 A, pp. 8-12, The National Archives, Kew, UK.

25 'Cork Street and Savile Row Area: Old Burlington Street', in *Survey of London: Volumes 31 and 32, St James Westminster*, Part 2, ed. F H W Sheppard (London, 1963), pp. 495-517.

26 Ambrose Vanderpoel, *History of Chatham New Jersey* (New York: Charles Francis, 1921), p. 449.

27 *The London Gazette*, 10 February 1781.

28 Thomas Simes, *The Regulator* (London, 1780), 120.

29 Ibid.

30 Thomas Simes, *A Military Guide for Young Officers* (London, 1776), p. 312.

31 Edward R. Curtis, *The Organization of the British Army in the American Revolution* (Gansvoort, NY: Corner House Historical Publications, 1998), 158.

32 Simes, *A Military Guide*, p. 314.

33 Sir Frederick William Hamilton, *The Origin and History of the First or Grenadier Guards*, 3 Vols. (London: J. Murray, 1874), Vol 2. p. 252.

34 Muster book of HMS Warwick, ADM 36/8657, The National Archives.

35 Orders and Instructions for His Majesty's 33d Regiment of Foot on their embarking for North America, 29 November 1775, WO 26/29 pp. 156-157, The National Archives; Orders for the 2d Battn of the 71st Regt on its embarkation for North America, 25 April 1776, Concord Free Public Library, Concord, MA.

36 *The American Rebellion: Sir Henry Clinton's Narrative of His Campaigns, 1775-1782*, ed. William B. Willcox (New Haven, CT: Yale University Press, 1954), p. 307.

37 *New York Gazette*, 2 July 1781.

38 *The American Rebellion: Sir Henry Clinton's Narrative*, p.532.

39 *New York Gazette*, 2 July 1781.

40 Charles Cornwallis to George Sackville Germain, 17 March 1781, *Colonial and State Records of North Carolina* Vol. 17 (Raleigh, NC: Josephus Daniels, 1899), 1002-1007.

41 Charles O'Hara to the Duke of Grafton 20 April 1781, "Letters of Charles O'Hara to the Duke of Grafton," ed. George C. Rogers, Jr., *South Carolina Historical Magazine*, v. 65 (1965): 178.

42 Johann Ewald, *Diary of the American War: A Hessian Journal*, ed. Joseph P. Tustin (New Haven, CT: Yale University Press, 1979), p. 314.

43 Ibid., p. 314

44 Alexander Leslie, to Cornwallis, 24 June 178, British Headquarters Papers, PRO 30/11/6/261-262, The National Archives.

45 Banastre Tarleton, *A History of the Campaigns of 1780 and 1781, in the Southern Provinces of North America* (Dublin: Colles, Exshaw, White, H. Whitestone, Burton, Byrne, Moore, Jones and Dornin, 1787), p. 310.

46 Notebook with daily account of operations against the Americans in the War of Independence, written by an officer of the 76th Regiment 1779-1781, GD146/18/6, Robertson of Kindeace papers, National Archives of Scotland.

47 Katherine Mayo in particular gives an elaborate description of the event that seems to be based entirely on the *Hibernian Magazine* account. Mayo, *General Washington's Dilemma*, p. 164.

48 *The American Revolution in North Carolina*, https://www.carolana.com/NC/Revolution/nc_camden_county_regiment.html.

49 Don N. Hagist, "Unpublished writings of Roger Lamb: Soldier in the American War of Independence," *Journal of the Society for Army Historical Research* 90, no. 362 (Summer 2012): p. 77.

50 Cornwallis to Sir Henry Clinton, October 20, 1781, in Tarleton, *A History of the Campaigns*, pp. 439-47.

51 *Bath Chronicle*, 15 August 1782.

52 Cornwallis to Sir Henry Clinton, October 20, 1781, in Tarleton, *A History of the Campaigns*, pp. 439-47.

53 Hugh Wodehouse Pearse, *Memoir of the life and military services of Viscount Lake: Baron Lake of Delhi and Laswaree, 1744-1808* (Edinburg: W. Blackwood and Sons, 1908), p. 64.

54 Moses Hazen to George Washington, 10 July 1782, *Founders Online*, National Archives, https://founders.archives.gov/documents/Washington/99-01-02-08886.

55 Richard Butler to Washington, 1 June 1782, *Founders Online*, National Archives, https://founders.archives.gov/documents/Washington/99-01-02-08581.

56 Ibid.

57 Washington to Butler, 10 June 1782, *The Writings of George Washington from the Original Manuscript Sources 1745-1799*, Vol. 24, ed. John C. Fitzpatrick (Washington, DC: Government Printing Office, 1938), p. 329.

58 General orders, 23 June 1782, ibid., pp. 373-74.

59 Washington to Benjamin Lincoln, 23 June 1782, ibid., pp. 374-75.

60 Hazen to Washington, 10 July 1782, *Founders Online*, National Archives, https://founders.archives.gov/documents/Washington/99-01-02-08886.

61 Ibid.

62 Butler to Washington, 11 July 1782, *Founders Online*, National Archives, https://founders.archives.gov/documents/Washington/99-01-02-08900.

63 Ibid.

64 Lincoln to Washington, 1 June 1782, *Founders Online*, National Archives, https://founders.archives.gov/documents/Washington/99-01-02-0880.

65 General orders, 21 November 1782, *The Writings of George Washington from the Original Manuscript Sources 1745-1799*, Vol. 25, ed. John C. Fitzpatrick (Washington, DC: Government Printing Office, 1938), pp. 366-67.

66 Washington to John Hanson, 20 April 1782, *Founders Online*, National Archives, https://founders.archives.gov/documents/Washington/99-01-02-08207.

67 Washington to Henry Clinton, 21 April 1782, *Founders Online*, National Archives, https://founders.archives.gov/documents/Washington/99-01-02-08216.

68 Washington to John Hanson, 20 April 1782, *Founders Online*, National Archives, https://founders.archives.gov/documents/Washington/99-01-02-08207.

69 Clinton to Washington, 25 April 1782, *Founders Online*, National Archives, https://founders.archives.gov/documents/Washington/99-01-02-08256.

70 Washington to Hazen, 3 May 1782, *Founders Online*, National Archives, https://founders.archives.gov/documents/Washington/99-01-02-08319.

71 Articles of Capitulation between Washington and Cornwallis, 19 October 1781, *Founders Online*, National Archives, https://founders.archives.gov/documents/Washington/99-01-02-07199.

72 Washington to Hazen, 18 May 1782, *Founders Online*, National Archives, https://founders.archives.gov/documents/Washington/99-01-02-08451.

73 Graham, *Memoir*, p. 80-81.

74 Martha Abel, "'Unfortunate': Lancaster, Pennsylvania, May 26–28, 1782," *The Journal of Lancaster County's Historical Society* Vol. 120 No. 3 (2019), pp. 97–105.

75 Graham, *Memoir*, p. 80-81.

76 Mayo, *General Washington's Dilemma*, pp. 263-68.

77 "Traditions of the American War of Independence," *United Service Magazine and Naval Military Journal* Part 3 (November 1834), p. 316. This account includes enough passages identical to Samuel Graham's *Memoir* that it was clearly written by Graham, perhaps as a draft of the *Memoir*.

78 Ibid.

79 Ibid.

80 Ibid.

81 James Gordon to Washington, 27 May 1782, *Founders Online*, National Archives, https://founders.archives.gov/documents/Washington/99-01-02-08531.

82 Mayo, *General Washington's Dilemma*, p. 139.

83 Hazen to Washington, 27 May 1782, *Founders Online*, National Archives, https://founders.archives.gov/documents/Washington/99-01-02-08533.

84 "Traditions of the American War of Independence," pp. 318-19. "Wayside Samaritans" is a term used by Katherine Mayo. Mayo, *General Washington's Dilemma*, p. 245.

85 Asgill to the Editor of the *New Haven Gazette*, 20 December 1786, see Appendix 3.

86 Asgill to Washington, 30 May 1782, *Founders Online*, National Archives, https://founders.archives.gov/documents/Washington/99-01-02-08562.

87 Graham, *Memoir*, p. 90.

88 "Traditions of the American War of Independence," p. 319.

89 Washington to Elias Dayton, 11 June 1782, *Founders Online*, National Archives, https://founders.archives.gov/documents/Washington/99-01-02-08661.

90 Ibid.

91 Dayton to Washington, 18 June 1782, *Founders Online*, National Archives, https://founders.archives.gov/documents/Washington/99-01-02-08719.

92 Trial of Richard Lippencott, WO 71/95 p. 320-408, The National Archives.

93 Washington to Dayton, 11 June 1782, *Founders Online*, National Archives, https://founders.archives.gov/documents/Washington/99-01-02-08661.

94 Asgill to the Editor of the *New Haven Gazette*, 20 December 1786, see Appendix 3.

95 "Day's Tavern" is shown prominently on the map "Road from near Morristown through Bottle Hill and Chatham towards Springfield. No 75, A / by Robert Erskine F.R.S. Geogr. A. U.S. and Assistants," New York Historical Society Digital Collections, https://digitalcollections.nyhistory.org/islandora/object/nyhs%3A243328.

96 Arthur D. Pierce, *Cradles of Revolt: The Taverns* (New Brunswick, NJ: Rutgers University Press, 1988), Chapter 7; Edmund B. Raftis, *Summit, New Jersey: From Poverty Hill to the Hill City* (Carolina, RI: Great Swamp Press, 1996), pp. 41-42.

97 Charles A. Philhower, *Brief history of Chatham Morris County, New Jersey* (New York: Lewis Historical Publishing Co, 1914).

98 Trial of Richard Lippencott, WO 71/95 p. 320-408, The National Archives.

99 Matthew H. Ward, "Joshua Huddy: the Scourge of New Jersey Loyalists," *Journal of the American Revolution*, 8 October 2018, https://allthingsliberty.com/2018/10/joshua-huddy-the-scourge-of-new-jersey-loyalists/

100 Asgill to the Editor of the *New Haven Gazette*, 20 December 1786; see Appendix 3.

101 Gordon to Sir Guy Carleton, 12 October 1783, British Headquarters Papers, PRO 30.55.83.89, The National Archives.

102 Asgill to the Editor of the *New Haven Gazette*, 20 December 1786, see Appendix 3.

103 Mayo, *General Washington's Dilemma*, p. 256.

104 "Traditions of the American War of Independence," p. 320.

105 *Kidd's Own Journal* (London), 24 April 1852, p. 265.

106 Vanderpoel, *History of Chatham New Jersey*, p. 435.

107 George James Ludlow to Washington, 31 July 1782, *Founders Online*, National Archives, https://founders.archives.gov/documents/Washington/99-01-02-08978.

108 Washington to Ludlow, 5 August 1782, *Founders Online*, National Archives, https://founders.archives.gov/documents/Washington/99-01-02-09033; Ludlow to Washington, 14 August 1782, *Founders Online*, National Archives, https://founders.archives.gov/documents/Washington/99-01-02-09133.

109 See, for example, *The London Chronicle*, 16 July 1782.

110 Thomas Townshend to Guy Carleton, 10 July 1782, CO 5/106/019, Colonial Office Papers, The National Archives.

111 James Jay to Washington, 19 July 1782, *Founders Online*, National Archives, https://founders.archives.gov/documents/Washington/99-01-02-08934.

112 *Aris's Birmingham Gazette*, 22 July 1782.

113 Carleton to Washington, 7 July 1782, *Founders Online*, National Archives, https://founders.archives.gov/documents/Washington/99-01-02-08851.

114 Carleton to Washington, 25 July 1782, *Founders Online*, National Archives, https://founders.archives.gov/documents/Washington/99-01-02-08948.

115 James Madison to Edmund Randolph, 16 July 1782, *Founders Online*, National Archives, https://founders.archives.gov/documents/Madison/01-04-02-0195.

116 *The New York Packet*, 10 October 1782. This newspaper also reported Asgill's age as fifteen years.

117 Madison to Pendleton, 27 August 1782, *Founders Online*, National Archives, https://founders.archives.gov/documents/Madison/01-17-02-0324.

118 Washington to Dayton, 25 August 1782, *Founders Online*, National Archives, https://founders.archives.gov/documents/Washington/99-01-02-09235.

119 Asgill to Dayton, 6 September 1782, *Proceedings of the New Jersey Historical Society* Vol. 8, 1856-1859 (Newark, NJ: printed at the Daily Advertiser Office, 1859), p. 198. The opening of this letter indicates that it duplicated a letter of 5 September sent by a different carrier.

120 Asgill to Dayton, 12 September 1782, ibid., p. 199.

121 Asgill to Dayton, 6 September 1782, ibid., p. 198.

122 Asgill to Dayton, 12 September 1782, ibid., p. 199.

123 Roger Lamb, *An original and authentic journal of occurences during the late American war, from its commencement to the year 1783* (Dublin: Wilkinson & Courtney, 1809), p. 424.

124 Robert Montgomery, *The Hawks of Hawk-Hollow. A Tradition of Pennsylvania* (Philadelphia: Carey, Lee & Blanchard, 1835), p. 199.

125 Asgill to Washington, 27 September 1782, *Founders Online*, National Archives, https://founders.archives.gov/documents/Washington/99-01-02-09600.

126 *The New York Packet*, 10 October 1782.

127 Asgill to Washington, 18 October 1782, *Founders Online*, National Archives, https://founders.archives.gov/documents/Washington/99-01-02-09752.

128 *Bath Chronicle*, 8 August 1782.

129 "Drawing lots for death," *The Daily Commonwealth* (Topeka, Kansas), 6 July 1887.

130 Anne Ammundsen, "Miss Asgill's Minuet," *Metropolitan: The Journal of the London Westminster & Middlesex Family History Society* Vol. 7 No. 2 (2021): 88–89. The minuet is catalogued at: "Miss Asgill's minuet [Música notada]". Europeana. And on "Charles Asgill's" YouTube.

131 Thomas Townshend to Guy Carleton, 10 July 1782, CO 5/106/019, The National Archives.

132 See Appendix Five.

133 Mayo, *General Washington's Dilemma*, p. 229.

134 comte de Vergennes to Washington, 29 July 1782, *Founders Online*, National Archives, https://founders.archives.gov/documents/Washington/99-01-02-08956.

135 Sir Thomas Townshend to Carleton, 14 August 1782, CO5/106/021, The National Archives.

136 *The Life, Public Services, Addresses and Letters of Elias Boudinot, LL.D, President of the Continental Congress*, ed. J.J. Boudinot, Vol. I (Boston and New York: Houghton, Mifflin and Company, 1896), pp. 249-51.

137 Ibid. Lady Asgill's letter was to the Foreign Minister, not the Queen.

138 Washington to Asgill, 13 November 1782, *Founders Online*, National Archives, https://founders.archives.gov/documents/Washington/99-01-02-09931.

139 *Kidd's Own Journal* (London), 24 April 1852, p. 265.

140 Washington to Asgill, 13 November 1782, *Founders Online*, National Archives, https://founders.archives.gov/documents/Washington/99-01-02-09931.

141 Washington to Hazen, 4 June 1782, *Founders Online*, National Archives, https://founders.archives.gov/documents/Washington/99-01-02-08600.

142 Washington to Hazen, 18 May 1782, *Founders Online*, National Archives, https://founders.archives.gov/documents/Washington/99-01-02-08451.

143 For more on the management of prisoners of war, see T. Cole Jones, *Prisoners of War and the Politics of Vengeance in the American Revolution* (University of Pennsylvania Press, 2019).

144 Henry Sinclair to Cortland Skinner, 27 May 1782, British Headquarters Papers, PRO 30/55/4675, The National Archives.

145 John Schaak to Carleton, 15 November 1782, British Headquarters Papers, PRO 30/55/54/66, The National Archives.

146 Carleton to Washington, 11 December 1782," *Founders Online*, National Archives, https://founders.archives.gov/documents/Washington/99-01-02-10174.

147 Carleton to Townshend, 17 January 1783, British Headquarters Papers, PRO 30/55/60/68, The National Archives.

148 A List of the Officers of the Army, 1783, WO 65/33 p. 128, The National Archives.

149 Washington to Vergennes, 21 November 1782, *Founders Online*, National Archives, https://founders.archives.gov/documents/Washington/99-01-02-10002.

150 Asgill to the Editor of the *New Haven Gazette*, 20 December 1786, see Appendix 3.

151 Gordon to Carleton, [17?] November 1782, British Headquarters Papers, PRO 30/55/55/98, The National Archives.

152 List of Quarters occupied by His Excellency the Commander in Chief, and his Suit, Sir Henry Clinton Papers, Vol. 267 Book C, William L. Clements library.

153 Asgill to the Editor of the *New Haven Gazette*, 20 December 1786, see Appendix 3.

154 *Hibernian Journal*, 23 December 1782; the account was carried in many British newspapers.

155 *Northampton Mercury*, 23 December 1782; *Saunders's News-letter* (London), 24 December 1782.

156 Mayo, *General Washington's Dilemma*, p. 245

157 *Reading Mercury*, 30 December 1782.

158 *The Times* (London), 3 February 1785.

159 Lady Asgill quoted this passage from Mayer's book, which had moved her greatly.

160 Sarah Asgill always signed as "Theresa" when writing to people she did not know; or, as in the case of writing to Vergennes, she used the French variant, Thérèse. Lady Colvile, Zelie Isabelle Richaud de Preville Colvile, *History of the Colvile Family* (Edinburg: Morison and Gibb, 1896), p. 195.

161 John Adams to Robert R. Livingston, 23 June 1783, Founders Online, National Archives, https://founders.archives.gov/documents/Adams/06-15-02-0025.

162 *London Gazette*, 14–18 May; Mackesy, *War for America*, p. 456–459

163 *Reading Mercury*, 30 December 1782, under a London 26 December dateline.

164 See Appendix 5.

165 See, for example, the London dateline in the *Northampton Mercury*, 9 September 1782.

166 See, for example, London dateline in the *Northampton Mercury*, 30 December 1782.

167 Asgill to Washington, 20 December 1786, see Appendix 3.

168 Ibid.

169 Statement of the Service of Lieutenant General Sir Charles Asgill Bart. Colonel of the 11th Regiment of Foot, WO 25/744 A, pp. 8-12, The National Archives, Kew, UK.

170 *Derby Mercury*, 30 October 1783.

171 *Calcutta Gazette*, 24 June 1784, translating from *Mercure de France*, 3 November 1783.

172 *Hereford Journal*, 13 November 1783.

173 *Calcutta Gazette*, 24 June 1784, translating from *Mercure de France*, 3 November 1783.

174 Antony Brett-James, *General Graham* (London: Macmillan, 1959), p. 14.

175 "George Montagu, 4th Duke of Manchester papers 1779-1788," William L. Clements Library, https://quod.lib.umich.edu/c/clementsead/umich-wcl-M-1536man.

176 Graham, *Memoir*, p. 74.

177 Ibid., pp. 104, 105.

178 Ibid., p. 107.

179 Asgill to the Editor of the *New Haven Gazette*, 20 December 1786, see Appendix 3.

180 *The New York Gazette and Weekly Mercury*, 20 October 1783; Graham, *Memoir*, p. 107.

181 *The Observer*, 29 May 1938.

182 Mayo, *General Washington's Dilemma*, p. 12.

183 E-mail, Trinity Church to the author, 22 August 2021.

184 *Hereford Journal*, 25 December 1783.

185 Graham, *Memoir*, p. 107.

186 A roll listing 348 men of the 1st Foot Guards serving in the Brigade of Guards in America includes 324 born in English counties; the remaining 22 were from Scotland (8), Wales (7), Ireland (5) and Minorca (2). Ledger of Accounts of the 1st Company, 1st Battalion, Brigade of Guards, November 1777 - March 1779, National Archives and Records Administration, Washington, DC.

187 https://www.geograph.org.uk/photo/199150.

188 *The Life, Public Services, Addresses and Letters of Elias Boudinot*, p. 251.

189 Thomas Jones, *History of New York During the Revolutionary War: And of the Leading Events in the Other Colonies at that Period*, Vol. 2 (New York: The New York Historical Society, 1879), p. 485.

190 *The New Haven Gazette and Connecticut Magazine*, 16 November 1786. See Appendix Two.

191 *The Writings of George Washington*, vol. X, (1782-1785), ed. W. C. Ford (New York: G. P. Putnam's Sons, 1891).

192 Asgill to the Editor of the *New Haven Gazette*, 20 December 1786, see Appendix 3.

193 Washington to David Humphreys, 26 December 1786," *Founders Online*, National Archives, https://founders.archives.gov/documents/Washington/04-04-02-0408.

194 Michael Abel, "The Asgill Letter," *The Journal of Lancaster County's Historical Society* Vol. 120 No. 3 (2019): 135.

195 Garden, *Anecdotes*, pp. 30-31.

196 Ibid., p. 33.

197 Ibid., pp. 33-34.

198 Ibid. pp. 38-39.

199 Mayo, *General Washington's Dilemma*, p. 265.

200 Washington to Edmund Pendleton, 22 January 1795, *Founders Online*, National Archives, https://founders.archives.gov/documents/Washington/05-17-02-0282.

201 *Leeds Intelligencer*, 14 June 1785.

202 *Ipswich Journal*, 19 March 1785.

203 *The Times* (London), 26 January 1788.

204 *The Gentleman's Magazine*, September 1788, p. 841.

205 *Bath Chronicle*, 2 October 1788; *Kentish Gazette*, 30 September 1788.

206 *The Gentleman's Magazine*, September 1788, p. 841.

207 Will of Sir Charles Asgill of Cork Street Burlington Gardens, Middlesex, PROB 11/1169/313, The National Archives.

208 Ibid.

209 F. G. Hilton Price, *A Handbook of London Bankers, with some account of their predecessors the Early Goldsmiths* (London: The Leadenhall Press, 1891), p. 123.

210 *Old Bailey Proceedings Online* (www.oldbaileyonline.org), September 1818, trial of William Goodman (t17881820-62).

211 William Knollis to Countess Banbury, 5 September 1790, 1M44/93/9, Hampshire Local Studies Collection, Hampshire Record Office. Asgill was 28 at this time.

212 Asgill v. Chandless b.r, C 12/167/4, The National Archives. While this very large document gives a full account of the hearing, the ruling (outcome) was not, apparently, available.

213 *The Times* (London), 16 September 1802.

214 Brian Masters, *Georgiana Duchess of Devonshire* (London: Hamish Hamilton, 1981).

215 *Kentish Gazette*, 12 March 1790.

216 George Turner and James Russell, *Reports of Cases Argued and Determined in the High Court of Chancery* (London, 1823), pp. 265–269.

217 *Kidd's Own Journal* (London), 24 April 1852, p. 265.

218 *The New Annual Register for the Year 1790*, p. 52.

219 *Survey of London: Volumes 29 and 30, St James Westminster*, Part 1, ed. F.H.W. Sheppard (London: London City Council, 1960), pp. 285–287.

220 Pictures and details of the house are in 'Duke of York Street' Survey of London: Volumes 29 and 30, St James Westminster, Part 1, ed. F.H.W. Sheppard (London: London City Council, 1960), pp. 285–287, https://www.british-history.ac.uk/survey-london/vols29-30/pt1/pp285-287.

221 "Remarkable Domestic Events, August 1791," The Historical Magazine, Vol. 3 No. 34 (August 1791), p. 302.

222 William McKay, W. Roberts, *John Hoppner RA* (London: P. & D. Colnaghi & Co and George Bell and Sons, 1909), p. xxiii.

223 *The Times* (London), 25 November 1791.

224 Daphne du Maurier's novel *Mary Anne* (1954) is a fictionalised account of the real-life story of her great great grandmother, Mary Anne Clarke née Thompson.

225 *Old Bailey Proceedings Online* (www.oldbaileyonline.org), December 1791, trial of Sarah Paris (t17911207-38).

226 Graham, *Memoir*, Chapter 10.

227 *The Times* (London), 5 December 1794.

228 Statement of the Service of Lieutenant General Sir Charles Asgill Bart. Colonel of the 11th Regiment of Foot, WO 25/744 A, pp. 8-12, The National Archives.

229 *The Times* (London), 10 March 1795.

230 Hamilton, *The Origin and History of the First or Grenadier Guards*, Vol 2 p. 322.

231 John Philippart, *The Royal Military Calendar* Vol. I (London: J. Valpy, 1815), pp. 115–16.

232 *The London Gazette*, 9 January 1798.

233 Steve Brown, British Regiments and the Men Who Led Them 1793-1815: 1st Regiment of Foot Guards, https://www.napoleon-series.org/military-info/organization/Britain/Infantry/Regiments/c_1stFootGuards.html.

234 Statement of the Service of Lieutenant General Sir Charles Asgill Bart. Colonel of the 11th Regiment of Foot, WO 25/744 A, pp. 8-12, The National Archives.

235 Liam Chambers, *The 1798 rebellion in north Leinster* (Dublin: Four Courts Press, 2003), pp. 122–36.

236 Richard Musgrave, *Memoirs of the different rebellions in Ireland* (Dublin: Robert Parchbank, 1801).

237 Ibid.

238 *Norwich Packet*, 14 August, 1798.

239 Turtle Bunbury Irish Histories, The Magistrate: Benjamin Bunbury (1751-1823) of Moyle & Killerig, https://turtlebunbury.com/document/benjamin-bunbury-magistrate/.

240 Robert Hugh Wallace, "Royal Downshire Militia. Extracts from Order Books, &c. (Continued)," *Ulster Journal of Archaeology*, Vol. 13, No. 2 (1907): pp. 59–69.

241 This is an amalgamation of snippets found in the diaries of several ladies who were Sophia's contemporaries.

242 *A Family Chronicle derived from Notes and Letters selected by Barbarina, the Hon. Lady Grey*, ed. Gertrude Lyster (London: John Murray, 2020), pp. 16-17. Barbarina, Lady Grey was the granddaughter of Barbarina Ogle.

243 *Voice of Rebellion - Carlow in 1798: The Autobiography of William Farrell*, ed. Roger Joseph McHugh (Dublin: Browne and Nolan, 1949), pp. 200-04.

244 *Byron's "Corbeau Blanc" The Life and Letters of Lady Melbourne*, ed. Jonathan David Gross (Houston, TX: Rice University Press, 1997), p. 278.

245 "Poor Mary Maher," *The Celt* (Dublin), 28 November 1857.

246 Sold at auction 23 June 2009, Lot 134, James Adam & Sons Ltd. Fine Art Auctioneers, Dublin.

247 Burwash, William & Sibley, Richard, London, 1807 Lot 174: George III Silver Presentation Hot Water Urn, 25 February 2006.

248 Thomas Pakenham, *The Year of Liberty* (London: Hodder and Stoughton, 1969), p. 284.

249 "Sliabh na mban – Slievenamon," https://irishpage.com/songs/slevmoan.html. And on "Charles Asgill's" YouTube.

250 *The London Gazette*, 13 May 1800.

251 Statement of the Service of Lieutenant General Sir Charles Asgill Bart. Colonel of the 11th Regiment of Foot, WO 25/744 A, pp. 8-12, The National Archives.

252 Alan Counihan, "A Cautionary Tail," *Local History Review* Vol. 18 (2013), pp. 87–90.

253 Brett-James, *General Graham*, p. 109.

254 Ibid., p. 108.

255 Ibid., p. 110.

256 Ibid.

257 *Granville Leveson Gower, Lord Granville Leveson Gower (first earl Granville): private correspondence, 1781 to 1821* ed. Castalia, countess Granville Vol. 1 (New York: E. P. Dutton, 1916), p. 368.

258 Ibid., pp. 335-36.

259 Gregory William Eardley-Twisleton-Fiennes, 14th Baron Saye and Sele (1769–1844).

260 *The Times* (London), 16 September 1802. The location was most likely to have been at Robert Barfield's Library which, like Stephen Nuckell's Library, offered dining and entertainment space, but Nuckell's was closed on Tuesdays. Both venues were popular with the aristocracy. Nuckell's still exists in the form of the Charles Dickens Pub, and parts of Barfield's are now restaurants in Albion Street. Barfield House, his home, is still nearby. Descendants of the Lords Saye and Sele confirm no knowledge of their ancestor owning property in Broadstairs.

261 Statement of the Service of Lieutenant General Sir Charles Asgill Bart. Colonel of the 11th Regiment of Foot, WO 25/744 A, pp. 8-12, The National Archives.

262 Patrick Geoghegan, *Robert Emmet: A Life* (Montreal: McGill-Queen's University Press, 2002), pp. 3-20.

263 *The London Gazette*, 12 January 1805.

264 Charles James, *A collection of the charges, opinions and sentences of general courts martial* (London: T. Egerton, 1820), p. 211-15.

265 Ibid.

266 Ibid.

267 Ibid.

268 Ibid.

269 *The London Gazette*, 15 February 1806.

270 *The London Gazette*, 1 November 1806.

271 *The London Gazette*, 3 March 1807.

272 R.E.R. Robinson, *The Bloody Eleventh: History of the Devonshire Regiment*, Vol. 1, 1685-1815 (Exeter: Devonshire and Dorset Regiment, 1988).

273 Arthur Bryant, *Years of Victory 1802-1812* (London: Collins, 1944), 312.

274 Richard Cannon, *Historical Record of the 11th Foot or North Devon Regiment of Foot* (London: Parker, Furnivall and Parker, 1845), p. 54.

275 *Freemans Journal* (Dublin), 26 May 1807.

276 Charles Lever, *Jack Hinton, The Guardsman* (London, George Routledge & Sons Ltd., 1843), p. 67.

277 Brett-James, *General Graham*, p. 115.

278 Ibid. p. 116.

279 Ibid.

280 Lady Morgan (Sydney) nee Owenson, *The Book Of The Boudoir* Vol. 1 (New York: Harper, 1829), p. 62.

281 Ibid., pp. 62-63.

282 *Supplementary Despatches, Correspondence, and Memoranda of Field Marshal Arthur Duke of Wellington, K.G.* Vol. 6 (London: John Murray, 1860), pp. 195-96.

283 "Poor Mary Maher," *The Celt* (Dublin), 28 November 1857.

284 Pam Moore, "Sir Charles Ogle: A Worthy Admiral," *The Portsmouth Papers*, 53 (1988).

285 *Debrett's Baronetage of England* (London: William Pickering, 1840), p. 412.

286 Marilyn Butler, *The Works of Maria Edgeworth*, Part I. (Routledge, 2019).

287 Maria Edgeworth to C. Sneyd, 3 December 1809, The Edgeworth Papers, National Library of Ireland.

288 James Taylor, *The Great Historic Families of Scotland* (London: J. S. Virtue, 1887).

289 Royal Collection Trust: Lady Frances Shelley (1787-1873) 1865-70 RCIN 2809171

290 Lady Frances Shelley (Winckley), *The Diary of Frances Lady Shelley 1787–1817*, Vol. 1, ed. Richard Edgcumbe (New York: C. Scribner's, 1912), pp. 49-50.

291 Brett-James, *General Graham*, p. 260.

292 Cecil Faber Aspinall-Oglander, *Freshly remembered: the story of Thomas Graham, Lord Lynedoch* (London: Hogarth Press, 1956), p. 83.

293 Ibid.

294 Correspondence of Sir Thomas Graham, National Library of Scotland. See Appendix Eight.

295 Charles Loraine Smith, Lady Asgill's Setter Dog – Original 1809 graphite drawing. Somerset & Wood.

296 *Women's Literary Networks and Romanticism A Tribe of Authoresses*, ed. Andrew O. Winckles and Angela Rehbein, (Liverpool: Liverpool University Press, 2017), p. 213-14.

297 Mary Tighe, "The Shawl's Petition, to Lady Asgill. 1772-1810," *Psyche, With Other Poems* (London: Longman, Hurst, Rees, Orme, and Brown, Paternoster Row, 1811), pp. 271-73.

298 HRH Frederick Duke of York to Lieutenant General Sir Charles Asgill, 3 January 1812, and Asgill's reply to Colonel John McMahon, 11 January 1812, Georgian Papers Online, Royal Collection Trust.

299 Earl of Hopetoun, *Oxford Dictionary of National Biography* (Oxford: Oxford University Press, 2004).

300 Digby Smith, *The Napoleonic Wars Data Book* (Barnsley, UK: Greenhill, 1998), p. 524.

301 HRH Frederick Duke of York to Lieutenant General Sir Charles Asgill, 3 January 1812, and Asgill's reply to Colonel John McMahon, 11 January 1812.

302 Philippart, *The Royal Military Calendar*, pp. 115–16.

303 *The Dispatches of Field Marshal the Duke of Wellington*, ed. John Gurwood, Vol. 7 (London: John Murray, 1852), p. 53.

304 *Freemans Journal*, 16 January 1812.

305 Brett-James, *General Graham*, p. 249.

306 'Duke of York Street', in *Survey of London: Volumes 29 and 30, St James Westminster, Part 1*, ed. F H W Sheppard (London, 1960), pp. 285-287.

307 Stephen Hoare, *Palaces of Power: The Birth and Evolution of London's Clubland* (Cheltenham, UK: History Press, 2019). The Secretary of the present-day Cavalry and Guards Club, David Cowdery, confirms that the club was formed in 1810 at 88 St. James's Street, and that original membership documents no longer exist. Legend has it that the three founding members were as listed. It can, therefore, only be assumed that this is the club Asgill joined.

308 Aspinall-Oglander, *Freshly remembered*, pp. 273-78. It is possible that Asgill joined The United Service Club instead (or even as well), which was formed in 1815, by Lord Lynedoch. By 1816, the Club had moved into its first premises in Albermarle Street. Three years later, in 1819, it moved to Charles Street. The club closed in 1978.

309 I am grateful to Martin Cherry, Librarian at the Museum of Freemasonry, London, for this information. The Star & Garter pub was at either 44 or 51 Pall Mall (there being two pubs of the same name in Pall Mall at that time).

310 *The Examiner* (London), 1 August 1813.

311 *The London Gazette*, 7 June 1814.

312 Aspinall-Oglander, *Freshly remembered*, p. 82.

313 Moore, "Sir Charles Ogle: A Worthy Admiral."

314 Brett-James, *General Graham*, p. 263.

315 Ibid., pp. 285-86.

316 PROB 11/1581/298: Will of Dame Sarah Asgill, Widow of Saint Marylebone, Middlesex, 21 June 1816, The National Archives.

317 Walter Sichel, *Sheridan: from New and Original Material; Including a Manuscript Diary by Georgiana Duchess of Devonshire* (London: Constable & company ltd., 1909), p. 386.

318 *The Times* (London), 15 August 1815.

319 Brett-James, *General Graham*, p. 314.

320 Ibid.

321 William A. Shaw, *The Knights of England* Vol. 1 (London: Sherratt and Hughes, 1906), p. 449.

322 *Survey of London: Volumes 29 and 30, St James Westminster*, Part 1, pp. 285–87.

323 The Thomas Phillips RA portrait of Charles Asgill, painted in 1822 and exhibited at the Royal Academy of Arts London that year, is listed in the National Portrait Gallery, London catalogue for the 1822 as: "107 Portrait of Gen. Sir Charles Asgill, Bart. G.C.G.O. T Phillips. R.A."

324 Charles Asgill to his tailor on Saturday 9 February. The Morgan Library & Museum. MA 489.81. Purchased by Pierpont Morgan, 1907.

325 *New Times* (London), 17 March, 9 April, 21 April 1823; *Morning Herald* (London), 2 April, 7 April 1823.

326 *New Times* (London), 22 April 1823.

327 *Morning Post* (London), 23 April 1823.

328 Will of Sir Charles Asgill of York Street Saint James's Square in the City of Westminster, Middlesex, PROB 11/1674/133, The National Archives.

329 See, for example, *The Gentleman's Magazine* (September 1823), pp. 274–75.

330 Aspinall-Oglander, *Freshly remembered*, p. 283.

331 Charles Ogle to Thomas Phillips, 23 October 1823, Morgan Library, New York.

332 Will of Sir Charles Ogle, PROB 11/1719/211, The National Archives.

333 Taylor, *The Great Historic Families of Scotland*.

334 Aspinall-Oglander, *Freshly remembered*, p. 82.

335 Ibid., pp. 82-83.

336 Brett-James, *General Graham*, p. 97.

337 Ibid., pp.137-39.

338 Aspinall-Oglander, *Freshly remembered*, p. 83.

339 Brett-James, *General Graham*, p. 51.

340 Aspinall-Oglander, *Freshly remembered*, p. 84.

341 Ibid., p. 135.

342 Ibid., p. 136.

343 Brett-James, *General Graham*, p. 97.

344 Aspinall-Oglander, *Freshly remembered*, p. 137.

345 Ibid., p. 142.

346 Ibid., p. 143.

347 Ibid.

348 Brett-James, *General Graham*, p. 114.

349 Aspinall-Oglander, *Freshly remembered*, p. 145.

350 Ibid., p. 161.

351 Brett-James, *General Graham*, p. 238.

352 Shorncliffe Camp, Ross Barracks. Shorncliffe Camp: https://shorncliffe-trust.org.uk/shorncliffe-camp-2/.

353 Philip R. Grant, *A Peer Among Princes* (Barnsley, UK: Pen & Sword Military, 2019).

354 Aspinall-Oglander, *Freshly remembered*, p. 178-86.

355 Aspinall-Oglander, *Freshly remembered*, p. 163

356 Ibid., p. 202.

357 Ibid., p. 220.

358 Ibid., pp. 231-36.

359 Brett-James, *General Graham*, p. 197.

360 Ibid., p. 113.

361 Ibid., p. 219.

362 Ibid., p. 251.

363 Aspinall-Oglander, *Freshly remembered*, pp. 243-45. The French line translates to "may he accept the assurance of my friendship and my regret."

364 Brett-James, *General Graham*, p. 21. Charles Loraine Smith (the son of a Northumberland baronet) was a friend of Graham's, so it is not surprising that he should sketch Sophia's dog.

365 Aspinall-Oglander, *Freshly remembered*, p. 280.

366 Ibid.

367 Ibid., p. 281.

368 Brett-James, *General Graham*, p. 313n1. Sir Charles Easy is a character in Colley Cibber's play *The Careless Husband* (1704), one who hated trouble.

369 *The London Gazette*, 7 May 1814.

370 Brett-James, *General Graham*, p. 301.

371 *Dictionary of National Biography*, 1885-1900, Volume 22.

372 Brett-James, *General Graham*, p. 301.

373 Aspinall-Oglander, *Freshly remembered*, p. 284.

374 Ibid., pp. 299-300.

375 Brett-James, *General Graham*, p. 336.

376 Taylor, *The Great Historic Families of Scotland*.

377 *The London Gazette*, 30 August 1823.

378 Brett-James, *General Graham*, p. 314-15.

379 Will of Sir Charles Asgill of York Street Saint James's Square in the City of Westminster, Middlesex, PROB 11/1674/133, The National Archives.

380 Ibid.

381 Ibid.

382 Graham, *Memoir*, p. 301.

383 Aspinall-Oglander, *Freshly remembered*, p. 287.

384 *Oxford Dictionary of National Biography*.

385 Aspinall-Oglander, *Freshly remembered*, p. 293.

386 *The Times* (London), 3 September 1839.

387 Courtesy of Joanne McLean, a Canadian descendant of Caroline (Asgill) Legge.

388 Courtesy of Joanne McLean. *Gloucester Journal*, 22 February 1840.

389 Courtesy of Joanne McLean. *Gloucestershire Chronicle*, 29 February 1840.

390 According to *Wilts & Gloucestershire Standard*, 3 June 1845, Caroline died at the home of her son-in-law, Joseph Bennett, whose late wife was Caroline's daughter, Theresa Frances Legge (who had predeceased her mother on 26 June 1844).

391 *New Times* (London), 8 September 1823.

392 Ibid.

393 *Bath Chronicle*, 2 October 1823.

394 Catherine Curzon, *The Scandal of George III's Court* (Barnsley, UK: Pen & Sword Military, 2018), p. 89.

395 Ibid., p. 127.

396 Cicero W. Harris, "Early Times in Granville County," transcribed by Tina Tarlton Smith, *The Torch Light*, 19 March 1878.

397 *Blackburn Standard*, 22 February 1854.

398 *Bath Chronicle*, 2 October 1823.

399 John Philippart, *The Royal Military Calendar* Vol. II (London: J. Valpy, 1820), pp. 6–7.

400 *Office-Holders in Modern Britain: Volume 11 (Revised), Court Officers, 1660-1837*, ed. R O Bucholz (London: University of London, 2006).

401 "Manners, Robert (1758-1823), of Bloxholm, Lincs," *The History of Parliament: the House of Commons 1754-1790*, ed. L. Namier (Martlesham, Suffolk: Boydell & Brewer, 1964).

402 *Kelly's Court Directory*, 1866-1867.

403 *Kentish Gazette*, 1 March 1870.

404 A record of this burial can be found at https://kentarchaeology.org.uk/research/monumental-inscriptions/loose.

405 Will of Mary Ann Mansel, Gentlewoman of Loose, Kent, PROB 11/2210/479, The National Archives.

406 Caregiver is the only possibility as given on: Acronym Definition https://acronyms.thefreedictionary.com/CG.

407 His death was recorded in the England and Wales, Civil Registration Index: 1837-1983 Record, Maidstone, Kent, Volume 2a, p. 392.

408 Will of Robert Manners, General in His Majesty's Service of Bloxholm, Lincolnshire, PROB 11/1673/144, The National Archives.

409 *The Gentleman's Magazine, and Historical Chronicle*, Volume 93, Part I 1823, p. 567.

410 *Old Bailey Proceedings Online* (www.oldbaileyonline.org), September 1818, trial of Richard Wheeler (t18180909-200).

411 Insured: Mary Ann Mansell, 15 Park Place South near The Man in the Moon Chelsea, MS 11936/488/978737, London Metropolitan City Archives.

412 *The Gentleman's Magazine*, September 1823, p. 274–75.

413 Mansel v Mansel, Bill and two answers, C 13/305/15, The National Archives.

414 Mansel letters.

415 1851 census, HO 107-1616 - page 19 folio 440, The National Archives.

416 *Parliamentary Papers*, Vol. 36 (London: HM Stationers, 1854), entry for 23 July 1847.

417 "History," All Saint's Church, Loose, UK.

418 Will of Sir Charles Asgill of York Street Saint James's Square in the City of Westminster, Middlesex, PROB 11/1674/133, The National Archives. William George Maton M.D. (31 January 1774 – 30 March 1835) was an English physician, a society doctor who became associated with the British royal family. In 1816 he was appointed physician extraordinary to Queen Charlotte, and in 1820 attended the Duke of Kent in his last illness. He afterwards became physician to the Duchess of Kent, and to the infant Princess Victoria (later to become Queen Victoria). He published on natural history and antiquarian topics.

[419] Philhower, *Brief history of Chatham Morris County, New Jersey*, p. 9, 21, 22.

420 Peter R. Henriques, "'Unfortunate': The Asgill Affair and George Washington's Self-Created Dilemma," *First and Always: A New Portrait of George Washington* (Charlottesville, VA: University of Virginia Press, 2020).

421 John H. Wheeler, *Reminiscences and memoirs of North Carolina and eminent North Carolinians* (Columbus, OH: Columbus print works, 1884), p. 178.

422 Edward Everett Hale, *The life of George Washington, studied anew* (New York and London: G.P. Putnam's Sons, 1888), p. 255.

423 *London Daily News*, 26 February 1903.

424 Hale, Edward Everett (1822-1909), Harvard Square Library, 13 November 2013.

425 Anne Ammundsen, "Saving Captain Asgill," *History Today*, Vol. 61 No. 12 (December 2011): 38-43.

426 Washington to Hazen, 18 May 1782, *Founders Online*, National Archives, https://founders.archives.gov/documents/Washington/99-01-02-08451.

427 This letter is referred to, above, in the newspaper article. It was the first letter Asgill wrote to Washington, to which he never received a reply.

428 Asgill to Washington, 30 May 1782, *Founders Online,* National Archives, https://founders.archives.gov/documents/Washington/99-01-02-08562.

429 Asgill to Washington, 27 September 1782, *Founders Online*, National Archives, https://founders.archives.gov/documents/Washington/99-01-02-09600. Washington's reply to this letter was included in his Papers. But he makes no reference whatsoever to the treatment Asgill has told him about.

430 Asgill to Washington, 18 October 1782, *Founders Online*, National Archives, https://founders.archives.gov/documents/Washington/99-01-02-09752.

431 "The Conduct of General Washington, respecting the Confinement of Capt. Asgill, placed in its true Point of Light," *The New Haven Gazette and Connecticut Magazine*, 16 November 1786. The intention, by publishing this correspondence, was to protect Washington's reputation, not just in 1786, but for posterity. In so doing, it damaged Asgill's reputation, irreparably, for two and a half centuries. This newspaper article was transcribed by Anne Ammundsen from a microfiche copy provided by Professor Robert Tombs. Italicising is as it was in that newspaper.

432 The date, May 17th, cannot be right – lots had not yet been drawn on 17 May. Clearly, it was 17 June. Asgill had been housed in Dayton's quarters for approximately ten days on 17 June, and he had been treated well by Dayton. Asgill's letter of the "30th of last month," has been reproduced at the commencement of these Appendices, and was written on 30 May 1782 from Philadelphia – just three days after the lots were drawn. He was on his journey to Chatham, so had no experience, then, of how he would be treated at when he was confined at the tavern there.

433 A copy of Asgill's letter was acquired, at my request, by Professor Robert Tombs of Cambridge University. The original letterbook was sold, in 2008, to a private collector. It was first published in The Journal of Lancaster County's Historical Society Vol. 120, No. 3 (Winter 2019), pp 135-141. What is given here is a replica of Lancaster's professionally transcribed copy, first published in Lancaster. Asgill included a number of footnotes

referring to enclosures which are no longer with the letter. Asgill has made an error in stating he was referring to Washington's papers in the 24 August edition of the New Haven Gazette and Connecticut Magazine; he meant to say the 16 November 1786 edition. That newspaper was the first to publish Washington's Papers, but others followed in 1787.

434 *Journal of Lancaster County's Historical Society* Vol. 120 No. 3 (2019) p.171.

435 Sotheby's, New York City: *Epistolary Correspondence between General Washington and Captn Asgill.*

436 "Ira Lipman, Security Man Who Spoke Out for Air Safety, Dies at 78," *The New York Times*, 29 September 2019.

437 Asgill to Washington, 30 May 1782, *Founders Online*, National Archives, https://founders.archives.gov/documents/Washington/99-01-02-08562.

438 Washington to Asgill, 13 November 1782, *Founders Online*, National Archives, https://founders.archives.gov/documents/Washington/99-01-02-09931.

439 Asgill to Washington, 18 October 1782, *Founders Online*, National Archives, https://founders.archives.gov/documents/Washington/99-01-02-09752.

440 *Handbook of the Antiquarian Booksellers Association 2008-9*, ed. Michael Silverman.

441 "Source record number: 727-535": Rare Book Hub.

442 "Washington, George – Asgill Affair: A fascinating manuscript titled Epistolary Correspondence between General Washington and Captain Asgill," William Doyle Galleries, https://doyle.com/auctions/22sc02-selections-private-collection-barbara-and-ira-lipman/catalogue/134-washington#lot-134.

443 Mayo, *General Washington's Dilemma*, pp. 263-68. This letter does not appear in the New York, Harcourt, Brace & Co., 1938 edition.

444 Ibid., pp. 229-30.

445 I purchased a copy of this play from New York. It was delightful to find, inside, transcriptions, in French, of all the correspondence given on these pages. The play had been owned by James Lorimer Graham (1835 – 1876), an American who had lived and studied in France for several years. He purchased his copy from W.H Huntington, Eight Rue de Boursault, Paris, France.

446 While Clinton was still in office when Joshua Huddy was murdered, by the time Asgill had been handed the "Unfortunate" lot, in May 1782, Sir Guy Carlton had replaced Clinton as Commander in Chief of British forces in North America. Brilly creates an inaccuracy, in a letter designed to address Le Barbier's inaccuracies.

447 Jean Louis Le Barbier to Washington, 4 March 1785, *Founders Online*, National Archives, https://founders.archives.gov/documents/Washington/04-02-02-0276;

Washington to Le Barbier, 25 September 1785, *Founders Online*, National Archives, https://founders.archives.gov/documents/Washington/04-03-02-0242.

448 Barbara Frances Tarling, *Representations of the American War of Independence in the late eighteenth-century English novel*. PhD thesis, 2010, The Open University, p. 165.

449 Correspondence of Sir Thomas Graham, National Library of Scotland.

450 Sophia's brother Sir Charles Ogle's home was at 42 Berkeley Square, so I think it is safe to say she was staying with him when writing to Thomas Graham.

451 These words were so important to Graham that he could not bring himself to destroy this one letter – ever.

452 Sophia's sister, Barbarina Wilmot.

453 Arabella, Barbarina Wilmot's daughter, only 3 years old at the time.

454 Richard Streatfield, who married Sophia's sister, Jane, in 1792.

455 It is not known who Mrs P and V are.

456 It is not known who St. C is.

457 Pizarro fulfilled anticipation by having a long run, and Sheridan's share of the profits was £15,000.

458 It is not known who Henry is. Sophia's younger brother, Edward Henry (1772-1791) had died before this letter was written, so it cannot be him.

459 Eton College kindly checked their records, and Georgina Robinson replied to my query: "Unfortunately, I have been unable to find a complete set of names of the Captains from this period. Our records on boys are lacking in detail from this time and the names of the Captains was not recorded in our records until the middle of the 19th Century."

Index

Aberystwith, Wales: 165, 168-170
Aboukir Bay, Egypt: 117, 152
Adams, John: 77
Arbuthnot, Alexander: 3, 37
Aretakis, Nick: 223
Articles of Capitulation: 1, 34-35, 40, 60, 204, 220, 224, 229, 232
Asgill, Amelia: 6, 14, 51, 57-59, 79-82, 91, 104, 143-144, 150, 157-158, 162, 178
Asgill, Annabella: 6
Asgill, Caroline: 6, 79, 104, 106-107, 143, 150, 163-164, 178, 203
Asgill, Charles, 1st Baronet: 6-10, 12, 14, 17, 51, 55-56, 57, 79, 101-104, 164, 196, 198, 206
Asgill, Charles, 2nd Baronet: 2-5, 10-15, 17, 18, 20-22, 23-26, 27-31 32-42, 44-56, 57-66, 67-76, 77-82, 86-87, 88-89, 91-100, 101-113, 114-123, 124-132, 133, 138, 139-150, 151-152, 154-155, 160, 161, 164-169, 173, 175-178, 180, 185, 187, 191-195, 196-201, 203, 204-206, 207-217, 218-227, 228-231, 232-233, 243-248, 253-256
Asgill, Charles (uncle to Charles Asgill, 1st Baronet): 6
Asgill, Henry: 6
Asgill, Jemima Sophia: 105, 107-109, 113, 117, 119-121, 124-126, 129-131, 133-138, 140-144, 147, 148, 150, 151-152, 154-160, 161, 192, 198, 203, 249-251
Asgill, Lydia: 6
Asgill, Maria: 6, 104, 106-107
Asgill, Sarah: 6, 10, 51, 57-60, 63, 74-76, 77, 79, 81-82, 83-85, 98, 103-104, 143-144, 158, 162, 232-233, 237-240, 246
Asgill, William Charles: 176-179, 180, 191, 252-256
Asgill House: 57, 73-74, 175-176, 180, 198
Asgill's Rant (song): 10
Aspinall-Oglander, Cecil: 152, 158
Balfour, Robert: xv, 85, 200
Barclay, David: 3, 36
Birch, Alan: xviii, 3, 72
Belson, Charles Philip: 127-128
Black Bear Tavern: 3-4, 38-39
Black Swamp, Battle of: 20-21
Boudinot, Elias: 62-64, 92
Bouverie, Arabella: see Ogle, Arabella
Bower, Meg: 191, 202
Brand, Barbarina: 117-118, 125-126, 133, 135, 141, 146, 152-155, 159-160, 162, 251
Brand, Thomas: 147, 162
Brilly, Elie: 10, 237-239
Broadway, Manhattan: 70-71
Brown, Liz: 198
Butler, Richard: 28-30
Callet, Antoine-François: 61
Carleton, Sir Guy: 45, 48, 52-53, 59-60, 62, 66, 68-70, 204, 212-213, 219, 229
Carlisle, Pennsylvania: 28-30, 72
Cavendish, Harriet: 125

Chambers, Liam: 114
Chandless, Thomas: 1-5
Charleston, South Carolina: 18
Chatham, New Jersey: 41, 44-56, 58, 62, 64-65, 67-69, 72, 78, 83
Childs, Charles: 180-190, 191-195, 200, 203
Childs, Elizabeth Mary Mansel: 184, 189
Childs, William Herbert: 184
Childs, Leigh: 184, 191
Clinton, Sir Henry: 33, 37, 45-46, 52, 228
Clarke, Mary Anne: 109
Colvile, Amelia: see Asgill, Amelia
Colvile, Sir Robert: 91, 162
Connah, Douglas John: 245
Continental Congress: 42, 50, 53-54, 56, 62-64, 69, 99, 217, 230, 243
Cornwallis, Charles: 17, 18-20, 22, 23-25, 40, 70, 88, 114, 204
Crowley, Dean: 191
Dashwood, Elizabeth: 82
Dayton, Elias: 44-46, 49, 54, 93, 98, 212-215, 219, 221-222
Delavoye, Alexander M.: 153
Dimma, Thomas: 183
Dobb's Ferry: 69, 214, 218, 225
Drummond, Charles Edouard: 184
Drummond-Murray, Peter: 176, 184
Duane, James: 62
Dublin Castle: 127, 131, 139, 140
Durham, Cornelius: 149
Edgeworth, Maria: 133
Edmed, Frederic Aylin: 184
Edmed, Jim: 175

Eld, George: 3, 36-37
Ellis, Henry: 124
Elphinstone, George: 18
Emmet, Robert: 114-115, 126-127
Falmouth, Cornwall: 73
Farrell, William: 119-120
Feiling, Keith: 87
Fellowes, Julian: 200
Flanders Campaign: 110-112
Fontainbleau, France: 79, 81, 247
Franklin, William: 47
Frederica, Duchess of York: 107, 109, 198
Frederick, Duke of York: 101-102, 105, 108, 110, 129, 139, 152, 173, 183
Gage, Charlotte: 124, 142
Gage, Deborah: xiv, 199
Garden, Alexander: 11-12, 96-99
Georgiana, Duchess of Devonshire: 105-106
Gilpin, Sawrey: 117-118, 133
Glover, Gareth: 156
Goodchild, Mary Ann: see Mansel, Mary Ann
Goodman, William: 104
Gordon, James: 3-4, 26, 27, 36-42, 44-48, 65-66, 69-70, 73, 83-90, 162, 197-198, 199-200, 203, 222, 229-231, 237-239
Gordon, Mary Ellen: 85
Gould, Charles: 57
Gower, Granville Leveson: 126
Graham, James Lorimer: 242-243
Graham, Robert: 130, 142, 158
Graham, Samuel: 3, 36, 83, 89, 110, 162-163
Graham, Thomas: 125-126, 129-130, 134-136, 141, 142-

Index

143, 144, 146, 151-160, 203, 249-251
Grasse, Admiral François Joseph Paul de: 77
Green, Captain: 73
Gregory, Isaac: 21, 200, 203
Gregory, John: 21, 201, 204
Greville, Henry: 2, 3, 5, 36-37, 39, 50, 228-231
Hale, Edward E.: 202
Hammond, Thelma Celeste: 173
Hauptfuhrer, Fred: 176
Hauzinger, Josef: 80
Hawthorn, William: 3, 37
Hayward, Elizabeth: 184, 186, 192
Hazen, Moses: 1, 3, 27-31, 34-41, 67, 95, 204, 211, 220, 228-230
Henrietta, Countess of Bessborough: 126
Henriques, Peter: 200
Hicks, Robert: 174, 201-202
Hope, John: 139
Hoppner, John: 107-108, 199
Huddy, Joshua: 32-34, 37, 40, 42, 45, 46-48, 67, 68, 78, 87, 204, 208, 211, 213-214, 219, 222, 224, 228, 242, 245, 248,
Hughes, William Carlyon: 73
Hulsberg, Henry: 13
Humphreys, David: 92, 95-96, 207-208, 220
Inglis, Bishop Charles: 87
Ingram, James: 3, 37
Jay, James: 50-51
Kemp's Landing, Virginia: 20-21
Kennedy, Vicky: 247
Kiffer, Selby: 223
Kilkenny, Ireland: 117, 119, 121-122, 124, 200
King George III: 2, 40, 59, 77, 101, 173, 174, 180, 183, 246
King George IV: 147, 173
King Louis XVI: 60, 63-64, 69, 72, 75, 79-81, 95, 110, 248
King, Turi: 178-179
Knollis, William: 105
Lacoste, Henri de: 241, 242
Lafayette, Gilbert du Motier, Marquis de: 79
Lake, Gerard: 18, 26
Lamb, Roger: 22
Lambe, Lawrence: 242
Lancaster, Pennsylvania: 4, 25, 36-40, 44, 72, 83, 197, 228
Lauzun, Armand Louis de Gontaut, duc de: 79
Lawrence, Sir Thomas: 102
Lay, Paul: xiv, 199
Le Barbier, Jean-Louis: 82, 234-240
Legge, Caroline: see Asgill, Caroline
Legge, Richard: 107, 143, 163-164
Legge, William: 163
Leutchford, Mary: 177, 180, 258
Lincoln, Benjamin: 25, 29-30, 230
Lipman, Ira: 223, 226
Lippincott, Richard: 45, 52, 61, 67, 68, 70, 231, 246-248
Livingston, Robert R.: 77
Lombard Street, London: 6, 9
Loose, Kent, UK: 184-189, 192
Luckcuck, Kathleen: 175-176, 180, 183-184, 186
Ludlow, George: 2, 3, 18, 36-37,

39, 46, 50, 212-213, 215-216, 230-231
Lynedoch, Lord: see Graham, Thomas
McCurdy, William: 27-31, 204
McMahon, John: 139
Madison, James: 52
Maher, Mary: 121
Manners, George: 192
Manners, Robert: 180, 182-185, 187-188, 191-193, 195
Mansel, Henry Edward: 188
Mansel, Herbert: 180-182, 185, 192-195
Mansel, Lucy: 180-181, 185, 192
Mansel, Mary Ann: 148, 180-188, 190, 191-195, 204
Marie Antoinette, Queen: 60, 63, 79-81, 95, 110, 219-220
Maton, George: 161-162, 194
Mayer, Charles-Joseph: 74
Mayo, Katherine: 73, 87, 99, 196, 242
Meigs, Josiah: 95-96, 207, 225
Middleton, Henry: 11, 97-98
Miles, Kristin: xv, xvi, 200
Miles, Lawford: 3, 36
Miss Asgill's Minuet (song): 58
Montagu, George: 81-82
Moore, John: 121, 151, 155-156
Morris-Jumel Mansion, Manhattan: 86-88
Morrison, Andrew: 176-178, 180
Morristown, New Jersey: 54, 72, 215
New Haven Gazette: 93-96, 196, 204, 207-217, 218, 225
New York, city: 18, 19, 22, 23-24, 45-46, 50, 54-55, 64-66, 68-70, 73, 83, 88, 89, 198, 200, 202, 207, 212, 214, 227, 229, 232
New York, state: 62
Newton, Henry James Hall: 173
Newton, Joyce: 175
Newton, Philip: 173
Nisbet, Mary Hamilton: 182-183, 192
Noel, Chris: 200
Norfolk, Virginia: 18, 20
O'Hara, Charles: 20, 25
Ogle, Arabella: 125, 141, 147, 154, 159-160, 162
Ogle, Barbarina: see Wilmot, Barbarina
Ogle, Sir Chaloner: 105, 107, 113, 117, 144, 146, 151
Ogle, Chaloner: 148-149
Ogle, Sir Charles: 107, 124, 142, 147, 148-149, 154, 161-162, 194
Ogle, Hester: 250
Ogle, Hester Thomas: 113, 117, 133, 151
Ogle, Jane: 125
Ogle, Jemima Sophia: see Asgill, Jemima Sophia
Ogle, Newton: 151
Ogle, Sophia: 142, 149
Ogle, Thomas: 117, 152
Old Burlington Street, London: 12-14
Paris, Sarah: 109-110
Perrin, James: 3, 18, 36-37
Philadelphia, Pennsylvania: 29, 34, 40-42, 72, 205, 213-214, 224, 227, 232, 234
Phillips, Thomas: 147-148, 198, 203
Portsmouth, Virginia: 18, 20, 72

Index

Pratviel, Sarah Theresa: see Asgill, Sarah
Reakes, Janet: 175
Reardon, Patrick: 127-128
Regiments, American
   2nd New Jersey Regiment: 44
   1st Pennsylvania Regiment: 27
   5th Pennsylvania Regiment: 28
Regiments, British
   Foot Guards, 1st Regiment: 2, 14-17, 18, 25, 36, 42, 79, 91, 103, 106, 112, 114, 124, 258
   Foot Guards, 2nd (Coldstream) Regiment: 36, 57
   Brigade of Guards: 15, 17, 18-21, 26, 37, 40, 62, 157
   5th West India Regiment: 128
   7th Regiment of Foot: 73
   9th Regiment of Foot: 22
   11th Regiment of Foot: 128-129, 131, 162, 164
   17th Regiment of Foot: 36
   23rd Regiment of Foot: 22, 24, 36
   26th Regiment of Foot: 37
   28th Regiment of Foot: 127-128
   33rd Regiment of Foot: 37
   46th Regiment of Foot: 124
   57th Regiment of Foot: 67
   58th Regiment of Foot: 117
   76th Regiment of Foot: 36
   80th Regiment of Foot: 4, 26, 37, 83, 86, 87, 225
   85th Regiment of Foot: 128
   90th Regiment of Foot: 155
Reynolds, Joshua: 106, 145
Richmond Place: see Asgill House
Rochambeau, Jean-Baptiste Donatien de Vimeur, Comte de: 60, 77, 91, 208-209, 218-219, 246
Romney, George: 146
Sage, Agnes Carr: 245
Sandy Hook, New Jersey: 67, 72, 73
Saumarez, Thomas de: 3, 36
Schaak, John: 67-68
Shelley, Lady Frances: 134-135, 158-159
Sheridan, Richard Brinsley: 144-145, 152, 157, 250
Sliabh na mBan (song): 123
Smith, Charles Loraine: 136, 138, 159, 200
Streatfield, Richard: 125
Sullivan, Arabella: see Wilmot, Arabella
Swallow, packet boat: 70-73
Swift (ship): 112
Talbot, Robert: 162
Tarleton, Banastre: 109-110
Taylor, James: 200
Taylor, Sir Robert: 6, 8-9, 57. 101-102
The Shawl's Petition, to Lady Asgill (poem): 136-138
Thomson, Charles: 70
Thruman, Alison: 200
Tietze, Mary Keim: 47
Tighe, Mary: 136-138
Tilghman, James: 207-209, 221
Timothy Day's Tavern: 46-49, 197
Tombs, Robert: xiii, 197
Townshend, Thomas: 59, 62
Trinity Church, Manhattan: 87-89, 197, 199
University of Göttingen: 12
Urwin, Gregory: 197

Vanderstegen, Elizabeth: 6
Vaughan-Newton, Kate: 246
Vere, Joseph: 6
Vergennes, Charles Gravier, comte de: 60-61, 68-69, 77, 79, 98, 220, 232-233, 246-248
Versailles: 81
Wade, George: 14
Warwick, HMS: 18
Washington, George: 1-2, 25, 28-31, 33-38, 40, 42, 44-47, 50-52, 56, 59-65, 67-70, 77-78, 87-88, 91-100, 202, 204-106, 207-217, 218-224, 228, 232, 240, 246
Wellesley, Sir Arthur: 131-132, 139-140, 147, 152, 158, 159, 159
Wellington, Duke of: see Wellesley, Arthur
Westminster School: 10-12, 96-97, 177
Wheeler, Richard: 191
White, Captain: 3
White, Philip: 32-33, 47, 228
Whitlocke, Bulstrode: 3, 37
William Pepys & Company: 6
Williamsburg, Virginia: 20, 21, 72
Wilmot, Arabella: 146, 154
Wilmot, Barbarina: see Barbarina Brand
Wilmot, Valentine Henry: 117-118, 125, 152, 154, 162
Winkett, Lucy: 199
Witz, Mr.: 3
Woburn Abbey: 134, 136, 152, 158, 204
Worthy Park House: 105, 107
York, Pennsylvania: 27-28, 36, 38, 39, 72, 228
York Street, London: 107, 141, 143, 147, 154, 162, 192-193, 197-198
Yorktown, Virginia: 1, 23-25, 35, 57, 60, 72, 77, 79
Young, Jennie: 176-177
Zoffany, Johan: 174